Katarzyna-Maria Bison

Calcareous dinoflagellates related to the Messinian Salinity Crisis

Katarzyna-Maria Bison

Calcareous dinoflagellates related to the Messinian Salinity Crisis

Their response and environmental evolution in the eastern (Cyprus) and central (Sicily) Mediterranean

Südwestdeutscher Verlag für Hochschulschriften

Impressum/Imprint (nur für Deutschland/only for Germany)
Bibliografische Information der Deutschen Nationalbibliothek: Die Deutsche Nationalbibliothek verzeichnet diese Publikation in der Deutschen Nationalbibliografie; detaillierte bibliografische Daten sind im Internet über http://dnb.d-nb.de abrufbar.
Alle in diesem Buch genannten Marken und Produktnamen unterliegen warenzeichen-, marken- oder patentrechtlichem Schutz bzw. sind Warenzeichen oder eingetragene Warenzeichen der jeweiligen Inhaber. Die Wiedergabe von Marken, Produktnamen, Gebrauchsnamen, Handelsnamen, Warenbezeichnungen u.s.w. in diesem Werk berechtigt auch ohne besondere Kennzeichnung nicht zu der Annahme, dass solche Namen im Sinne der Warenzeichen- und Markenschutzgesetzgebung als frei zu betrachten wären und daher von jedermann benutzt werden dürften.

Coverbild: www.ingimage.com

Verlag: Südwestdeutscher Verlag für Hochschulschriften GmbH & Co. KG
Dudweiler Landstr. 99, 66123 Saarbrücken, Deutschland
Telefon +49 681 37 20 271-1, Telefax +49 681 37 20 271-0
Email: info@svh-verlag.de

Approved by: Bremen, Universität Bremen, Diss., 2011

Herstellung in Deutschland:
Schaltungsdienst Lange o.H.G., Berlin
Books on Demand GmbH, Norderstedt
Reha GmbH, Saarbrücken
Amazon Distribution GmbH, Leipzig
ISBN: 978-3-8381-2832-0

Imprint (only for USA, GB)
Bibliographic information published by the Deutsche Nationalbibliothek: The Deutsche Nationalbibliothek lists this publication in the Deutsche Nationalbibliografie; detailed bibliographic data are available in the Internet at http://dnb.d-nb.de.
Any brand names and product names mentioned in this book are subject to trademark, brand or patent protection and are trademarks or registered trademarks of their respective holders. The use of brand names, product names, common names, trade names, product descriptions etc. even without a particular marking in this works is in no way to be construed to mean that such names may be regarded as unrestricted in respect of trademark and brand protection legislation and could thus be used by anyone.

Cover image: www.ingimage.com

Publisher: Südwestdeutscher Verlag für Hochschulschriften GmbH & Co. KG
Dudweiler Landstr. 99, 66123 Saarbrücken, Germany
Phone +49 681 37 20 271-1, Fax +49 681 37 20 271-0
Email: info@svh-verlag.de

Printed in the U.S.A.
Printed in the U.K. by (see last page)
ISBN: 978-3-8381-2832-0

Gutachter

Prof. Dr. Helmut Willems

Prof. Dr. Rüdiger Henrich

Tag des Kolloquiums

20. Mai 2011

„The whole is simpler than the sum of its parts."

Josiah Willard Gibbs (1839-1903)

Calcareous dinoflagellate cyst evolution and their response to the environmental changes related to the Messinian salinity crisis in the eastern (Cyprus) and central (Sicily) Mediterranean

Table of contents

Preface

This project has been funded by the DFG (German Research Foundation) as a single project (Wi 725/19-1/-2) with the title "Reaktionen kalkiger Dinoflagellaten auf drastische Umweltveränderungen und Stresssituationen im Zusammenhang mit der Salinitätskrise des Messinian Event." The project was developed and implemented as a cumulative dissertation under supervision of Prof. Dr. Helmut Willems and guidance of Dr. Gerard J.M. Versteegh and supported by cooperating groups from the Netherlands (University of Utrecht) and France (Muséum National d'Histoire Naturelle; Université de Paris-Sud).

The thesis comprises three first-authorship manuscripts which are already published (Chapter 2, 3) or submitted (Chapter 4) for publication in peer-reviewed international scientific journals. The manuscripts are preceded by a summary and a general introduction (Chapter 1) followed by conclusions and future perspectives (Chapter 5). References are given at the end of each chapter. Chapter 1 provides an introduction to the group of dinoflagellates in general and in particular to the calcareous cyst producing dinoflagellates implying its application, classification and methods used. In addition it provides background information relevant to this thesis regarding the Mediterranean Sea, the Messinian salinity crisis (MSC) and the study area.

Summary

Marine microorganisms and their fossil relics (microfossils) play an important role in inferring past climatic and environmental changes. Preserved in the sedimentary records of the oceans, each group of these microfossils and its individual species provide different information about the climatic and environmental conditions in the water-column during their life time (e.g. temperature, salinity, nutrients). Therefore, past climatic and environmental conditions can be reconstructed by the use of different types of microfossils as proxies. Calcareous cyst producing dinoflagellates (calcareous dinoflagellates) were shown to have a high proxy potential, since they are geographically widespread in marine environments, highly responsive to environmental changes, resistant to calcite dissolution, and have a long stratigraphic range. Thus, they provide a good fossil record. This fossil record reflects the environmental conditions in the surface waters during their life time. Previous research on modern and living calcareous dinoflagellates has notably increased our knowledge of their ecology and environmental relationships. Their temporal and spatial distribution is strongly correlated with the environmental parameters of the surface waters and certain species are excellent indicators for specific environmental conditions. Furthermore, several studies on Quaternary sediments have confirmed their usability for environmental reconstructions. However, these studies mainly focused on Quaternary records derived from surface sediments. Palaeoenvironmental reconstructions based on older fossil calcareous dinoflagellates, especially of Tertiary age and from the Mediterranean realm, are very rare. From the Mediterranean region, only a few and primarily descriptive studies have been carried out on fossil calcareous dinoflagellates from pre-Quaternary sedimentary sequences. Consequently, only little is known about their potential for environmental reconstructions from this region. A major reason for this gap is that studies on fossil (Mesozoic/Tertiary) material used to be based on a time consuming single specimen picking method and a subsequent semi-quantitative analysis under the scanning electron microscope (SEM). Therefore, the major aim of this thesis is to demonstrate the applicability of fossil (pre-Quaternary) calcareous dinoflagellates as a tool for palaeoenvironmental reconstructions and to expand the existing knowledge by applying a faster and competitive method.

One of the most attractive natural laboratories to study past environmental and climatic changes is the Mediterranean Sea. Due to its nearly enclosed nature and its special geographic position, it reacts very sensitively to climatic changes. Therefore, marine Mediterranean sedimentary sequences provide a unique archive of astronomically induced past climatic and hydrological variations. During the Neogene, the Mediterranean region underwent extreme geodynamic processes and environmental changes. One of the most outstanding events is the Messinian salinity crisis (MSC) at the end of the Messinian stage. Because of this event, large contrasts were established in the environmental conditions, spatially and temporally. Hence, this event is very well suited to prove the response of calcareous dinoflagellates to drastic environmental changes and their applicability for palaeoenvironmental reconstructions. This is a primary justification for the study of the evolution of calcareous dinoflagellates in relation to the environmental changes preceding and following the MSC on an east-west transect through the Mediterranean.

Herewith, we present the first detailed study on calcareous dinoflagellates that systematically investigated their evolution with respect to the MSC on selected land sections from the Mediterranean realm.

After the first pilot studies, it became obvious that we had to modify our initial approach. First, we had to reduce the range of our east-west transect and in return increase the number of samples in order to obtain long- and short-term trends. As the most suitable land sections for our study, we selected the Pissouri Basin on Cyprus (eastern Mediterranean) and the Caltanissetta Basin on Sicily (central Mediterranean). Both locations provide almost complete sedimentary successions of the stratigraphic intervals preceding and following the MSC. Second, we had to modify the existing method established for Quaternary calcareous dinoflagellate studies. For quantification, originally we wanted to use the faster polarised light microscope technique (PLM). However, this turned out to be not feasible for our sample material. Therefore, we had to modify/develop a new and competitive method for the quantification under the SEM.

Our first study focused on the evolution of the calcareous dinoflagellate assemblages in the Pissouri Basin prior to (upper Tortonian/pre-evaporitic Messinian) and immediately after (early Pliocene, 5.33 Ma) the MSC. The deposits from the Pissouri Basin represent an intermediate water depth of about 300 to 500 m. We analysed 41 samples from the marl layers only. Our results show that the late Miocene/early Pliocene calcareous dinoflagellate associations are very different compared to those of the Mediterranean today where *Thoracosphaera heimii* overwhelmingly dominates, followed by *Calciodinellum levantinum* in the western and *Lebessphaera urania* in the eastern basin. The upper Tortonian/lower Messinian assemblages are clearly dominated by *Calciodinellum albatrosianum*, indicating relatively stable, warm, and oligotrophic surface water conditions. A significant drop of *C. albatrosianum* at ~6.42 Ma initiated a phase of drastic and frequent changes in the surface water conditions, characterised by extreme shifts in species dominance and an overall decrease in total cyst concentrations. This phase continued up to shortly before (~5.98 Ma) the onset of the MSC (5.96 Ma). Three main environmental phases have been distinguished prior to the MSC, followed by a significant change in the environmental conditions immediately after the MSC (~5.33 Ma). In the first phase (upper Tortonian), *C. albatrosianum* clearly dominates the calcareous dinoflagellate associations with high cyst concentrations, indicating extremely oligotrophic and stable warm surface water conditions, associated with a warm and dry climate with low continental influence. However, two notable drops at about 7.53 Ma and 7.24 Ma in total cyst numbers indicate short-term cooling of the surface waters, associated with a slight increase of continental influence (*C. stella*/*L. granifera*) and salinity (*L. urania*). In the second phase (lower Messinian), *C. albatrosianum* still dominates although its total cyst abundance strongly decreased, indicating a decrease in the surface water temperature, again associated with higher continental influence (*C. stella*) and salinity (*L. urania*). These modifications during the lower Messinian interval possibly reflect the beginning of more unstable and restricted conditions caused by the MSC. With the third phase, a significant change in the surface water conditions occurred, indicating highly unstable and restricted conditions. *C. albatrosianum* for the first time lost its permanent dominant position. At the same time, other species indicative for enhanced continental influence temporarily increased. With the beginning of the Pliocene, the association again significantly changed. Specifically, *L. granifera* replaced *C. albatrosianum* as the dominant species.

In the second study, we investigated the first 100 kyrs of the earliest Pliocene (Zanclean) following the MSC, again on samples from the Pissouri Basin on Cyprus. A total of 14 samples of marine marls (Trubi facies) were analysed. The marine sedimentary record represents a water depth of at least 300 m. Based on the shifts in calcareous dinoflagellate dominance, we differentiated between three intervals. The first is strongly dominated by *L. granifera*. This indicates highly eutrophic surface water conditions, lower salinities and moderate temperatures. These conditions we related to increased continental runoff and upwelling driven by an estuarine circulation. The second interval starts with a significant drop of *L. granifera* and an increase of *C. albatrosianum* and *C. stella*, indicating decreased nutrient concentrations, raised salinities and increased surface water temperatures (SST). This change in the surface water conditions represents a general shift in the climatic and hydrologic situation towards more dry and somewhat warmer conditions, and the beginning of an anti-estuarine circulation. The third phase is initiated by a distinct drop of *C. stella*, associated with an increase of *L. urania* and *Rhabdothorax* spp. This in general indicates a further decrease of the nutrient concentrations in the surface waters, whereas the increase of *Rhabdothorax* spp. may be related to the normalisation of bottom water conditions (sedimentary resting cyst stage) and the beginning of a higher seasonality. At the end of this interval, *T. heimii* for the first time notably increased and became the dominant species. This third phase denotes the establishment of typical open marine, well stratified and oligotrophic surface water conditions, similar, but not yet identical to the conditions prevailing in the Mediterranean today.

In the third study from the central Mediterranean Caltanissetta Basin on Sicily, we again investigated the sedimentary records preceding (upper Tortonian/pre-evaporitic Messinian) and following the MSC (the first ~120 kyrs of the Pliocene). In order to study a deeper basin setting, we selected the Gibliscemi, Falconara (Tortonian/Messinian) and Eraclea Minoa section (early Pliocene) as a composite. The estimated water depth is about 1200 m. Altogether, we analysed 63 samples from the marl and sapropel layers to reveal short- and long-term environmental changes. We distinguished seven major environmental phases, of which three we assigned to the upper Tortonian/Messinian and four to the early Pliocene sequence. The first phase is characterised by highly oligotrophic conditions with fluctuations in SST and salinity as indicated by *C. albatrosianum*, *P. tuberosa* and *L. urania*. *C. albatrosianum* dominates the associations more or less distinctly, except for the uppermost sample. Here (~7.51 Ma), a significant drop of this species indicates a short-term cooling of the surface water. In the following interval, *C. albatrosianum* dominates again. However, some notable shifts in relative and absolute abundance and the temporal increase of other species, such as *L. granifera*, *P. parva* and *Rhabodothorx* spp., indicate fluctuations in SST, salinity, nutrient concentrations and continental influence. A major drop of *C. albatrosianum* at 7.17 Ma indicates a cooling of the surface water. The third phase started with a drastic and abrupt change (at ~6.78 Ma) in the dinoflagellate cyst association, marked by drastic shifts in species dominance, episodic exceptional peak occurrences of *T. heimii* and *C. albatrosianum* (e.g. at ~6.52 Ma and ~6.47 Ma) and barren samples (i.e. ~6.67 Ma and ~6.63 Ma), and the complete disappearance of calcareous dinoflagellates at about 6.0 Ma. With the onset of the Pliocene (5.33 Ma), fully marine conditions were established and *L. granifera* clearly replaced *C. albatrosianum*. It was initiated by a prominent peak occurrence of *L. granifera*, indicating relatively cool and highly eutrophic conditions in the surface water. To explain this, we suggest intensified continental runoff in combination with local upwelling, the latter activated by an early estuarine circulation system, and a cooler and more humid climate phase. After a transitional period of about 120 kyrs, *C. albatrosianum* again became dominant. This indicates a renewed change in the hydrological and climatic conditions, from eutrophic and somewhat cooler surface waters (humid and cool climate) towards warmer and oligotrophic conditions (warm and arid climate), most likely associated with a change from an estuarine to anti-estuarine system, similar to prevailing conditions today in the Mediterranean Sea. However, the terminal Pliocene dinoflagellate association still differs from that of the present Mediterranean. Some of the most common species in the Mediterranean Sea currently (*Thoracosphaera heimii, Lebessphaera urania, Calciodinellum levantinum*) are still strongly underrepresented or even missing *(Calciodinellum elongatum*). On the other hand *C. albatrosianum* – the dominating species of our final Pliocene record - today accounts for no more than 5% of the Mediterranean dinoflagellate associations. This suggests that water surface temperatures were still warmer and that seasonality was less pronounced.

The comparison of our data from Sicily and Cyprus has shown that the main environmental trends, as reflected by the dinoflagellate cyst record, are in general accordance. However, some obvious deviations in the distribution pattern indicate local differences in tectonic activity, hydrology and climate of the eastern and central Mediterranean basins.

This thesis proved the applicability of fossil calcareous dinoflagellates for palaeoenvironmental reconstructions of the Late Neogene Mediterranean. It enhances the knowledge about both the spatial and temporal evolution of Neogene calcareous dinoflagellates in the central and eastern Mediterranean realm. In addition this study extends our knowledge about the environmental conditions preceding and following the MSC. Furthermore, it provides new insight into the hydrological conditions prevailing during sapropel and marl formation in the Mediterranean. Therefore, Neogene calcareous dinoflagellates represent a valuable tool for future palaeoenvironmental studies.

Zusammenfassung

Bei der Entschlüsselung vergangener Klima- und Umweltveränderungen spielen marine Mikroorganismen und deren fossile Überreste (Mikrofossilien) eine wichtige Rolle. Die verschiedenen Mikrofossilgruppen und deren einzelne Arten liefern unterschiedliche Informationen über die klimatischen und ökologischen Bedingungen zum Zeitpunkt ihres Lebens in der Wassersäule (z.B. Temperatur, Salzgehalt, Nährstoffe). Durch die Verwendung verschiedener Mikrofossilien als Stellvertreter (Proxy) für bestimmte Umweltfaktoren, können diese für die Rekonstruktion der Umweltbedingungen in den Ozeanen der Vergangenheit herangezogen werden. Kalkige Dinoflagellatenzysten haben ihr Proxy-Potential bereits in verschiedenen Studien gezeigt. Sie sind in den Ozeanen weit verbreitet, reagieren sensitiv auf Änderungen ihrer Umgebung, haben eine hohe stratigraphische Reichweite und sind aufgrund ihrer Resistenz gegen Kalkauflösung gut in marinen Sedimenten erhalten und liefern somit eine beträchtliche Datenbasis. An dem Vorkommen der Dinoflagellaten in den Sedimenten können daher die Umweltbedingungen im Oberflächenwasser zu ihren Lebzeiten abgelesen werden. In den letzen Jahren durchgeführte Studien an modernen und lebenden kalkigen Dinoflagellaten haben unser Wissen über die Ökologie und Umweltbeziehungen der Dinoflagellaten merklich vergrößert. Ihre räumliche und zeitliche Verteilung korreliert mit den Umgebungsparametern des Oberflächenwassers und bestimmte Arten sind exzellente Indikatoren für spezifische Umweltbedingen. Mittlerweile haben diverse Studien an quartären Sedimenten ihre Nützlichkeit für die Paläo-Rekonstruktion bestätigt. Es muss jedoch angemerkt werden, dass diese Studien sich ausschließlich auf quartäre Oberflächensedimente fokussierten. Paläorekonstruktionen an älterem fossilem Material, insbesondere des Tertiärs und aus dem Mittelmeerraum, wurden nur selten durchgeführt. Bislang gibt es für den Mittelmeerraum nur ein paar wenige und primär beschreibende Studien über kalkige Dinoflagellaten an prä-quartären Sedimentabfolgen. Folglich ist wenig über ihr Potential für die Rekonstruktion der Umweltbedingen der älteren (z.B. Tertiär) geologischen Vergangenheit bekannt. Ein Grund dafür ist, dass das fossile Material mit einem zeitlich sehr aufwändigen Selektions-Verfahren (Auspicken einzelner Zysten) und nachfolgender semi-quantitativer Analyse unter dem Rasterelektronenmikroskop (REM) ausgewertet wurde. Ein wesentliches Ziel dieser Doktorarbeit ist daher die Überprüfung der Anwendbarkeit der fossilen prä-quartären kalkigen Dinoflagellaten für die Paläo-Rekonstruktion und die Erweiterung des bestehenden

Wissens auf Basis der Weiterentwicklung der etablierten Methodik hin zu einem schnelleren und konkurrenzfähigen quantitativen Auswerteverfahren.

Eines der attraktivsten natürlichen Laboratorien zum Studium vergangener Umweltbedingungen und klimatischer Änderungen ist das Mittelmeer bzw. der Mittelmeerraum. Aufgrund der geographischen Position und der annähernden Abgeschlossenheit reagiert das Mittelmeer sehr sensitiv auf klimatische Änderungen. Die mediterranen marinen Sedimentabfolgen stellen ein einzigartiges Archiv von astronomisch induzierten vergangenen Klimata und hydrologischen Variationen dar.

Während des Neogens war die Mittelmeerregion extremen geodynamischen Prozessen und klimatischen Veränderungen ausgesetzt. Eines der einschneidendsten Ereignisse ist die Salinitätskrise (MSC) des Mittelmeers am Ende des Messiniums. Im Verlauf dieses Ereignisses bildeten sich starke räumliche und zeitliche Kontraste in den Umweltbedingungen heraus. Daher ist die Salinitätskrise besonders gut geeignet, um die Auswirkung der drastischen Umweltveränderungen auf die Dinoflagellaten zu untersuchen und damit deren Potential für die Paläorekonstruktion zu überprüfen. Deshalb wollten wir die Evolution der kalkigen Dinoflagellaten im Bezug auf die Umweltveränderungen vor und nach der Salinitätskrise auf einem Ost-West Transekt durch das Mittelmeer untersuchen. Mit dieser Studie präsentieren wir zum ersten Mal eine detaillierte und systematische Untersuchung der Evolution der kalkigen Dinoflagellaten im Zusammenhang mit der Salinitätskrise an ausgewählten Landaufschlüssen des Mittelmeerraumes.

Nach den ersten Pilotstudien wurde offensichtlich, dass wir den initialen Ansatz modifizieren mussten. Zum einen mussten wir aufgrund des abzusehenden großen Aufwandes den Schnitt durch das Mittelmeer auf ein paar ausgewählte Lokalitäten reduzieren und dafür aber die Probenanzahl erhöhen, um langzeitige und kurzzeitige Trends besser erfassen zu können. Als am Besten geeignet für unsere Studien wählten wir Landaufschlüsse im Pissouri-Becken auf Zypern (östliches Mittelmeer) und im Caltanissetta-Becken auf Sizilien (zentrales Mittelmeer) aus. In beiden Lokalitäten befinden sich nahezu komplette und gut erschlossene Abfolgen des stratigraphischen Intervalls vor und nach der Salinitätskrise. Außerdem mussten wir die, für Studien an quartären kalkigen Dinoflagellaten etablierte Methodik modifizieren bzw. teilweise neu entwickeln. Eigentlich wollten wir für die Quantifizierung ebenfalls die, sich für das Quartär etablierte, Polarisationslichtmikroskopische (PLM) Methodik verwenden. Dieses Verfahren stellte sich aber als nicht geeignet für unser Probenmaterial heraus. Daher modifizierten bzw. entwickelten wir eine neue Methode für die effiziente Quantifizierung unter dem REM.

Unsere erste Studie befasste sich mit der Entwicklung der kalkigen Dinoflagellaten Assoziationen im Pissouri-Becken in der prä-evaporitischen Phase der Salinitätskrise und unmittelbar nach der Krise (frühes Pliozän, 5.33 Ma). Die Ablagerungen im Pissouri-Becken repräsentieren eine mittlere Wassertiefe von ca. 300 bis 500 Metern. Insgesamt untersuchten wir 41 Proben aus den Mergel-Lagen. Unsere Ergebnisse zeigen, dass sich die Dinoflagellaten Assoziationen des späten Miozäns bzw. frühen Pliozäns deutlich von denen des heutigen Mittelmeeres unterscheiden, in dem *Thoracosphaera heimii* klar die Kalkdinoflagellaten Assoziationen dominiert, gefolgt von *Calciodinellum levantinum* im westlichen und *Lebessphaera urania* im östlichen Becken. Die Zusammensetzung der Assoziationen des oberen Tortoniums und unteren Messiniums dagegen werden klar von *C. albatrosianum* dominiert, welches wir mit relativ warmen und stabilen Oberflächenwasser-Bedingungen in Verbindung gebracht haben. Ein starker Abfall von *C. albatrosianum* um 6.42 Ma initiierte dann eine Phase von starken und häufigen Änderungen in den Bedingungen des Oberflächenwassers, angezeigt durch extreme Verschiebungen in der Artendominanz und einem allgemeinen Abfall der Zystenhäufigkeiten. Diese Phase dauerte an bis kurz vor dem Anfang (~5.98 Ma) der Salinitätskrise. Insgesamt konnten wir drei Phasen bis zur eigentlichen Salinitätskrise ausmachen. Nach der Salinitätskrise (5.33 Ma) fand eine erneute deutliche Veränderung in den Umweltbedingungen statt. In der ersten Phase (oberes Tortonium) wurde die Dinoflagellaten Assoziation klar von *C. albatrosianum* dominiert, begleitet von hohen absoluten Häufigkeiten. Dieses zeigt extrem oligotrophe und stabile warme Bedingungen im Oberflächenwasser, welche mit einem warmen und trockenen Klima und geringen kontinentalen Einflüssen assoziiert werden können. Es gibt jedoch auch zwei merkbare Abfälle in den totalen Zystenkonzentrationen um ca. 7.53 Ma und 7.24 Ma, welche für kurzzeitige Abkühlung im Oberflächenwasser stehen, assoziiert mit einer Zunahme des kontinentalen Eintrags (*C. stella* / *L. granifera*) und der Salinität (*L. urania*). In der zweiten Phase (unteres Messiniums) war *C. albatrosianum* weiterhin dominant, wenn auch mit stark reduzierter Zystenkonzentration. Dieses steht für eine Abnahme der Wassertemperaturen begleitet von einem höheren kontinentalen Einfluss (*C. stella*) und höherer Salinität (*L. urania*). Diese Veränderungen stellen erste Anzeichen der Salinitätskrise dar. Die dritte Phase ist gekennzeichnet durch eine signifikante Änderung der Oberflächenwasserbedingungen aufgrund der, durch die Salinitätskrise verursachten. zunehmenden Instabilität und Einschränkung des mediterranen Raumes. *C. albatrosianum* verlor ihre bis dahin uneingeschränkte Dominanz wenn auch mit gelegentlichen Erholungen. Zur gleichen Zeit stiegen die Konzentrationen anderer Arten temporär an. Mit dem Beginn des Pliozäns veränderten sich die Dinoflagellaten Assoziationen erneut. Die Zystenkonzentrationen stiegen wieder an und *L. granifera* verdrängte jetzt *C. albatrosianum* als dominierende Art.

In der zweiten Studie untersuchten wir die ersten 100 ka des frühen Pliozäns (Zanklium) nach der Salinitätskrise, wiederum an Proben aus dem Pissouri-Becken auf Zypern. Insgesamt untersuchten wir 14 Proben aus den marinen Mergeln (Trubi Fazies). Die marinen Sedimente repräsentieren Wassertiefen von mindestens 300 Metern. Basierend auf den Verschiebungen der Dinoflagellaten Dominanzen konnten wir drei Intervalle differenzieren. Ersteres wird dominiert von *L. granifera,* welches hohe Eutrophierung des Oberflächenwassers, eine Verringerung der Salzkonzentration und kühlere Temperaturen reflektiert. Diese Bedingungen lassen sich mit vermehrtem kontinentalen Eintrag und Auftrieb, verursacht durch eine estuarine Zirkulation, assoziieren. Das zweite Interval startet mit einem starken Abfall von *L. granifera* und einer Zunahme von *C. albatrosianum* und *C. stella,* welches ein Zeichen für abnehmende Nährstoffkonzentration und angestiegener Salinität und Oberflächenwassertemperatur ist. Diese Veränderungen in den Oberflächenwasserbedingungen stellen eine generelle Verschiebung der klimatischen und hydrologischen Situation dar, hin zu trockeneren und etwas wärmeren Bedingungen und dem Beginn einer anti-estuarinen Zirkulation. Mit der dritten Phase fiel *C. stella* stark ab und *L. urania* und *Rhabdothorax* spp. stiegen in ihrer Konzentration an. Dieses reflektiert die weitere Abnahme der Nährstoffkonzentration im Oberflächenwasser, wobei der Anstieg von *Rhabdothorax* spp. vermutlich mit der Normalisierung der Bedingungen im Bodenwasser (Ausbildung von Ruhe-Zysten) und dem Beginn einer größeren Saisonalität in Verbindung gebracht werden kann. Am Ende dieses Intervalls stieg die Konzentration von *T. heimii* zum ersten Mal merklich an und sie wurde damit zur dominierenden Art in der Dinoflagellaten Assoziation. Mit dieser dritten Phase stellten sich typische offenmarine Bedingungen ein, mit gut stratifizierten und oligotrophen Oberflächenwasserbedingungen, die den heutigen Bedingungen im Mittelmeer nahe kommen.

In unserer dritten Studie untersuchten wir erneut die Sedimentabfolgen vor und nach der Salinitätskrise, diesmal aber an Proben aus dem Caltanissetta-Becken auf Sizilien. Die Ablagerungen der Gibliscemi, Falconara und Eraclea Sektion stellen Sedimente aus einem tiefen Becken mit Wassertiefen von ca. 1200 Metern dar. Insgesamt untersuchten wir 63 Proben aus Mergel und Sapropel Schichten, um kurzfristige und langfristige Änderungen zu erfassen. Wir konnten sieben wesentliche Phasen ausmachen. Drei davon haben wir dem oberen Tortonium / Messinium und vier dem frühem Pliozän zugerechnet. Die erste Phase ist charakterisiert durch hohe Nährstoffarmut mit Fluktuationen in der Oberflächenwassertemperatur und Salinität, angezeigt durch die Arten *C. albatrosianum*, *P. tuberosa* und *L. urania*. *C. albatrosianum* dominiert die Assoziation mehr oder weniger deutlich mit Ausnahme der jüngsten Probe. In dieser zeigt der Abfall dieser Spezies um ca. 7.51 Ma eine kurzfristige Abkühlung der Oberlächenwassertemperaturen an. In dem darauf folgenden Intervall dominierte *C. albatrosianum* wieder, was erneute relativ stabile warme und oligotrophe Bedingungen anzeigt. Dennoch sind die gelegentlichen, aber deutlichen Verschiebungen in den relativen und absoluten Häufigkeiten, sowie die temporären Anstiege anderer Arten wie *L. granifera*, *P. parva* und *Rhabdothorax* spp., ein Indiz auf Fluktuationen in den Oberflächenwassertemperaturen, den Nährstoffkonzentrationen und dem kontinentalen Einfluss in diesem Intervall. Der merkliche Abfall von *C. albatrosianum* um 7.17 Ma. steht dabei beispielhaft für kühlere Oberflächenwassertemperaturen, assoziiert mit einem stärkeren kontinentalen Einfluss. Die dritte Phase startet mit einer drastischen Veränderung der Dinoflagellaten Assoziation um ca. 6.78 Ma. Im Durchschnitt fiel *C. albatrosianum* stark ab und es gab starke Verschiebungen der Speziesdominanz, episodisches Auftreten von besonders großen Häufigkeiten von *T. heimii* und *C. albatrosianum* (z.B. um 6.52 Ma und 6.47 Ma), gelegentliche Proben ohne Zysten und schließlich das komplette Verschwinden kalkiger Dinoflagellaten gegen 6.0 Ma. In der ersten Phase des Pliozäns waren dann vollmarine Bedingungen wieder hergestellt und *L. granifera* verdrängte *C. albatrosianum* als dominierende Spezies, iniziiert durch einen markanten Anstieg dieser Art, assoziiert mit relativ kühlen und oligotrophen Bedingungen im Oberflächenwasser. Außerdem schlagen wir für diese Phase einen erhöhten kontinentalen Eintrag, in Kombination mit lokalem Auftrieb, vor, letzterer aktiviert durch eine frühe estuarine Zirkulation, und eine kühlere und etwas humidere klimatische Phase. Nach einer Übergangsphase von ca. 120 ka übernahm *C. albatrosianum* erneut die Vormachtstellung in der Zystenassoziation. Dieses zeigt den erneuten Übergang von eutrophen und kühleren (feuchtes und kaltes Klima) hin zu wärmeren und oligotrophen (warmes und arides Klima) Bedingungen. Dieser Übergang war vermutlich auch mit einem Wechsel von einer estuarinen hin zu einer anti-estuarinen Zirkulation verbunden, wie sie im heutigen Mittelmeer anzutreffen ist. Dennoch unterscheidet sich die Dinoflagellaten Assoziation noch deutlich von der des heutigen Mittelmeeres. Die im heutigen Mittelmeer am häufigsten auftretenden Dinoflagellaten Arten (*T. heimii*, *L. urania*, *C. levantinum)* sind stark unterrepräsentiert oder fehlen vollständig wie z.B. *C. elongatum*. Die dominierende Art unseres abschließenden Pliozän Intervalls, *C. albatrosianum*, dagegen macht heutzutage im Mittelmeer nicht mehr als 5% der Dinoflagellatenassoziation aus.

Der Vergleich unserer Daten von Sizilien und Zypern zeigt eine generelle Übereinstimmung der Trends. Wie auch immer, die offensichtlichen Abweichungen in den Verteilungsmustern stehen für lokale Unterschiede in der tektonischen Aktivität, Hydrologie und dem Klima, im Bereich des östlichen und westlichen Mittelmeerraumes.

Wir konnten mit dieser Dissertation die Anwendbarkeit fossiler kalkiger Dinoflagellaten für die Rekonstruktion vergangener Umweltbedingungen am Beispiel des neogenen Mittelmeeres verifizierten. Diese Arbeit hat unser Wissen sowohl über die räumliche als auch zeitliche Evolution neogener kalkiger Dinoflagellaten im zentralen und östlichen Becken des Mittelmeeres erweitert. Darüber hinaus konnten wir neue Erkenntnisse über die Umweltbedingungen vor und nach der Salinitätskrise gewinnen. Ebenfalls erhielten wir neue Einblicke in die hydrologischen Bedingungen zur Zeit der Sapropel und Mergel Bildung im mediterranen Raum. Abschließend stellen wir fest, dass neogene Kalkdinoflagellaten ein wertvolles Instrumentarium für weitere zukünftige Paläoumweltstudien darstellen.

Chapter 1

The present is the key to the past is the key to the future

" ...whereas all experiences are of the past, all decisions are about the future... it is the great task of human knowledge to bridge this gap and find those patterns in the past which can be projected into the future as realistic images..." (Kenneth Boulding, 1973)

Ongoing natural climate variability has affected the earth's (weather) system throughout its history. There are different ways to look at the earth system and to measure climate variability generating different worlds and pictures. An approximate realistic picture of the earth's climate history is made up of many individual findings derived from various disciplines of science. Any information is like a little puzzle piece ultimately resulting in a nearly complete picture, that at least is our aspiration.

Understanding past geologic and ecosystem processes may improve our ability to predict future effects of global climate and environmental change.

Micropalaeontology is a discipline which deals with the history of the earth's biosphere and physical environments based on microfossil composition in a stratigraphic sequence. Microfossils are integrated in many ways in geological and environmental data and significantly contribute to the reconstruction of past climate and environmental change. One group that was long time overlooked in this respect are calcareous dinoflagellates, although they are known for a long time (e.g. Deflandre, 1947; Wall and Dale, 1968; Wall et al., 1970). These long ranging taxa which are known at least from the Triassic onwards (Janofske, 1992) have left a tremendous fossil record in marine sediments (e.g. Bolli, 1974; Fütterer, 1977; Keupp, 1981; Willems, 1988; Janofske, 1992; Kohring, 1993; Kienel, 1994; Zügel, 1994; Hildebrand-Habel et al., 1999; Streng, et al., 2004) since they are very resistant against carbonate dissolution. Many Tertiary species are still extant and allow us to draw conclusions from their present habitat to that of the past (e.g. Bison et al., 2007; 2009). Today calcareous dinoflagellates occur in all marine environments and different taxa are associated with different environmental conditions in the surface water. Hence their fossil remains can provide important information that can contribute to the interpretation of past environmental settings.

The main goal of this project was to proof the potential of fossil calcareous dinoflagellates as quantitative indicators for past environmental conditions and changes. Therefore the response of calcareous dinoflagellates to environmental changes in relation to the Messinian salinity crisis (MSC) in the Mediterranean realm has been investigated. This is the first study that systematically investigated a set of stratigraphically important and ecologically relevant reference profiles of the Mediterranean realm with respect to the MSC. The MSC as a perfect example of drastic environmental changes in a short period of time and the nearly land-locked Mediterranean basin with its sensitivity to climatic and environmental changes make up an ideal scenario and setting to proof and confirm the usability of fossil calcareous dinoflagellates for palaeoenvironmental reconstructions.

Our approach contributes to both, the understanding of the MSC in the Mediterranean and the evolution of calcareous dinoflagellates in response to the accompanying environmental changes. Besides this, the project represents a case study to extent our knowledge of Mediterranean Neogene calcareous dinoflagellates and their applicability as a tool for palaeoenvironmental reconstructions.

Main objectives and outline

The major aim of this thesis is to proof the applicability of fossil calcareous dinoflagellates for palaeoenvironmental reconstructions and increase our knowledge of Neogene Mediterranean calcareous dinoflagellates. For this, calcareous dinoflagellates shall be applied for the reconstruction of the environmental conditions and possible trends and events prior to the MSC and just after the reestablishment of marine conditions in the central Mediterranean Caltanissetta Basin (Sicily) and in the eastern Mediterranean Pissouri Basin (Cyprus).

Therefore, this study uses the following approaches:

- Calcareous dinoflagellate cyst distribution pattern shall be analyzed on an east-west transect from land sections throughout the Mediterranean region and correlated to palaeoenvironmental changes related to the MSC.
- Revision and modification of the established methods for Quaternary studies for its applicability to fossil material.
- Comparison of the results with previous studies on Neogene and modern Mediterranean calcareous dinoflagellate assemblages.
- Survey available sample material of selected Neogene Mediterranean reference profiles from international cooperation partners.
- Collection of new sample material from qualified localities during field surveys.

This study attempts to answer the following questions that arise from it:

- How did the calcareous dinoflagellates respond to the environmental changes related to the MSC?
- What happened to them after the refilling of the Mediterranean basin after the MSC?
- Did the prior Messinian dinoflagellate association re-establish?
- How do modern Mediterranean calcareous dinoflagellate assemblages compare to those of the studied stratigraphic interval?
- Are there differences in the spatial and temporal evolution related to the MSC?
- Can calcareous dinoflagellates be used as indicators for past environmental changes?
- Which modifications are necessary to establish fossil calcareous dinoflagellates as a competitive and reliable tool for palaeoenvironmental reconstructions?

- Are there key species which are indicative for specific environmental conditions?

The results and palaeoenvironmental interpretations of the calcareous dinoflagellate evolution during the time preceding (latest Miocene) and following (earliest Pliocene) the MSC in the eastern (Cyprus) and central (Sicily) Mediterranean area are presented as three manuscripts in the Chapters 2, 3 and 4.

Manuscript 1 – *Calcareous dinoflagellate turnover in relation to the Messinian salinity crisis in the eastern Mediterranean Pissouri Basin, Cyprus (Bison et al., 2007; Journal of Micropalaeontology)*

In the first manuscript we focused primarily on the evolution of the calcareous dinoflagellates in direction to the MSC and immediately after the refilling in the eastern Mediterranean (Cyprus).

Manuscript 2 – *Palaeoenvironmental changes of the early Pliocene (Zanclean) in the eastern Mediterranean Pissouri Basin (Cyprus) evidenced from calcareous dinoflagellate cyst assemblages (Bison et al., 2009; Marine Micropaleontology)*

The second manuscript represents a continuation of the first study. Here we investigated the calcareous dinoflagellate evolution during the first ~100 ka after the MSC.

Manuscript 3 - *Calcareous dinoflagellate cyst distribution and their environmental implications preceding and following the Messinian salinity crisis in the Caltanissetta Basin, Sicily (Bison & Willems; to be submitted for publication in Palaeogeography, Palaeoclimatology, Palaeoecology)*

In the third manuscript we investigated the evolution of the calcareous dinoflagellates in relation to the environmental changes preceding and following the MSC in the central Mediterranean Caltanissetta Basin (Sicily). The results have been compared with those from the Pissouri Basin to verify whether the pattern observed on our easternmost succession remain valid westwards.

Dinoflagellates (Dinophyceae)

"The great majority of the plants in the ocean are various types of planktonic, unicellular algae, collectively called phytoplankton." (Carol M. Lalli &Timothy R. Parsons, 2001)

Dinoflagellates (Division Dinoflagellata (Bütschli, 1885) Fensome et al., 1993), are a large group of microscopic, predominantly unicellular, flagellated eukaryotic, aquatic organisms, comprising a huge number of "algal" species of various shapes and sizes (e.g. Evitt, 1985; Taylor and Pollingher, 1987; Kohring, et al., 2005; Matthiessen et al., 2005). Their most common size range is between 20 to 200 μm. They are geographically widespread and occur from equatorial to polar latitudes (e.g. Gilbert and Clark, 1983; Stover et al., 1996; Maret and Zonneveld, 2003; Vink, 2004; Zonneveld et al., 2005; Rochon, 2009). They can be found in almost all types of aquatic environments, but with a majority (90%) in the marine coastal ones (Dale, 1992; Fensome et al., 1993; Taylor et al., 2008). Their spatial and temporal distribution in aquatic systems is attributed to a large number of physiological and biological parameters (e.g. light, temperature, nutrients, salinity, competition, turbulence) (Vink, 2004). Together with coccolithophores and diatoms they are the most important primary producers in the oceans (Parsons et al., 1984; Brasier, 1985; Taylor and Pollingher, 1987; Bujak and Brinkhuis, 1998; Rochon, 2009).

Currently, about 2000 extent and 2000 fossil species of dinoflagellates are known (Fensome et al., 1993). They follow different feeding strategies, from phototrophy and heterotrophy to mixotrophy (e.g. Dale, 1983; Parsons et al., 1984; Gaines and Elbrächter, 1987, 1991; Head, 1996; Schnepf and Elbrächter, 1992, 1999; Stoecker, 1999; Stickney et al., 2000; Matthiessen et al., 2005), which are again divided into several subunits (e.g. symbiotic, parasitic). About half of the extant species are heterotrophic or mixotrophic (Stoecker, 1999), while the other half is phototrophic (e.g. Parsons et al., 1984; Siano and Montresor, 2005). Although light is the main limiting factor, for phototrophic species only view dinoflagellate species are strict autotrophs, and most (i.e. auxotroph) need organic components such as vitamins or trace elements (e.g. iron) to grow (e.g. Taylor and Pollingher, 1987; Gaines and Elbrächter, 1987; Schnepf and Elbrächter, 1992). Actually many dinoflagellates are mixotrophs, capable to be autotrophic and heterotrophic at the same time (e.g. Sander, 1991; Stickney et al., 2000; Saldarriaga et al., 2001; Matthiessen et al., 2005).

Dinoflagellates typically occur as free-swimming motile mostly non-fossilisable stages, besides a potentially fossilisable non-motile stage. The vast majority of the fossil dinoflagellate record consists of cysts. Apart from their high morphological variability, dinoflagellates generally share a common anatomical pattern during at least one stage of their life cycle (Fig. 1). The motile stage generally shows a characteristic flagella arrangement, with a transversal flagellum that encircles the cell body, and perpendicular to it, a longitudinal flagellum (Fig. 1). The transversal flagellum is usually situated within an equatorial or slightly descending groove, called the girdle or cingulum. It runs to the left and usually encircles the cell completely. The proximal part of the longitudinal flagellum lies freely in another groove, termed the sulcus. The movement of the flagella causes the cell to migrate vertically in a helical path through the water column (Taylor and Pollingher, 1987; Fenchel, 2001). This character enables the motile cell to optimize their position in a stratified water column to a certain extent (Matthiessen et al., 2005; Clegg et al., 2007; Smayda, 2010). In contrast to a common believe that turbulences are unfavourable for dinoflagellate species (e.g. Wendler et al., 2002a, b; Vink, 2004) several authors (e.g. Smayda, 2000, 2002, 2010; Smayda and Reynolds, 2001; Sullivan and Swift, 2003; Sullivan et al., 2003) reviewed this paradigm, concluding that some dinoflagellates may cope well with turbulence. Laboratory experiments by Sullivan and Swift (2003) and Sullivan et al. (2003) showed that growth rates may be reduced at high turbulence for some species, but others were unaffected or even showed higher growth rates.

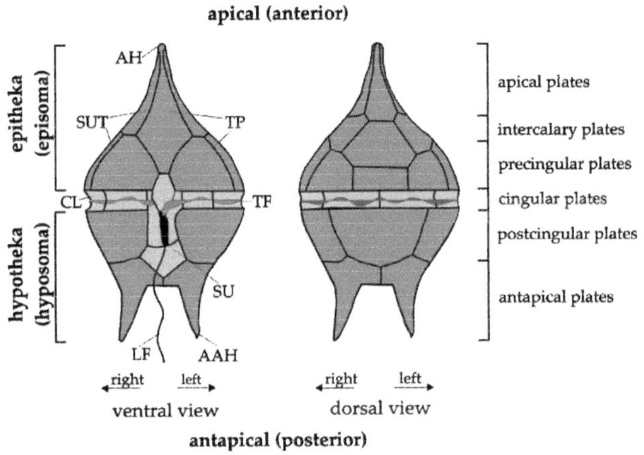

Fig. 1: Principle morphological features and terminology of a thecate motile cell of a peridinialean dinoflagellate (modified after Fensome et al., 1996).

Most dinoflagellates have a complex multi-staged life cycle (Evit, 1985; Burkholder et al., 2001 and citations herein; Fensome et al., 1996, 2003) (Fig. 2, 3). About 10 to 15% of the living dinoflagellates have a non-motile stage in their life cycle where they form highly resistant fossilisable cysts (resting cysts), which are organic-walled, calcareous or seldom siliceous (e.g. Wall and Dale, 1968; Taylor, 1987; Fensome et al., 1993, 1996). This study concentrates on the calcareous cyst producing dinoflagellates (calcareous dinoflagellates).

Calcareous dinoflagellates

Calcareous dinoflagellates are primarily phototrophic planktonic organisms and thus live in the photic zone of the oceans where sufficient light is available for photosynthesis and growth (e.g. Tangen et al., 1982; Binder and Anderson, 1987; Montresor et al., 1994). They are widespread in all marine environments, from open marine to coastal, and from sub-polar to tropic regions (e.g. Dale, 1992; Montresor et al., 1994, 1998, 2003; Wendler et al., 2002a, b; Vink, 2004; Zonneveld, et al., 2005), but most are found in higher latitudes where they can contribute considerably to the total carbonate flux to the sea floor (Dale, 1992). High abundances were also observed in neritic areas, both in temperate and tropical (e.g. Wall and Dale, 1968b; Dale, 1992; Montresor et al., 1994; 1998, 2003).

After coccospheres, calcareous dinoflagellates are the second most calcareous phytoplankton group and form an essential part of the first link in the marine trophic level, initially transferring light energy to chemical energy (photosynthesis). Thus, the majority of marine life depends on this energy transfer. Like most of the primary producer they relate to the environmental conditions in the surface waters, biotic and abiotic. Various studies about their distribution in modern oceans in fact have shown this (e.g. Dale, 1992 a, b; Höll et al., 1998; Esper et al., 2004; Vink et al., 2001; Vink, 2004; Wendler et al., 2002a, b, c; Zonneveld, 2003; Meier and Willems, 2003; Richter et al., 2007). They have proved to be rather resistant against carbonate dissolution (Karwath, 2000; Vink, 2001; Baumann, 2003; Esper et al., 2004; Zonneveld, 2000, 2004; Meier et al., 2004; Zonneveld et al., 2001, 2005) and are therefore generally well preserved in fossil sediments. Hence, due to their extensive fossil record (e.g. Keupp, 1981; 1987, 1991; Kohring, 1993a; Kienel, 1994; Hildebrand-Habel, et al., 1999; Dias-Brito, 2000; Streng et al., 2004a) they are apparently well suited for palaeoenvironmental reconstructions.

About 260 fossil species (morphotypes) (Fensome and Williams, 2004) and about 30 extant species (Vink, 2004; Zonneveld et al., 2005) have been described so far. Cyst size shows a great variety (intra- and extraspecific) with a range between about 7 and 50 µm (Zonneveld et al., 2005) for extant species. In fossil records the cyst size is generally larger than today and most commonly ranges between e.g. 40 and 100 µm (Dias-Brito, 2000) or from 20 to 75 μm (Wendler and Willems, 2001) in the Mesozoic. A general drop in cyst sizes of calcareous dinoflagellate species in the lower Cretaceous was observed by Wendler et al. (2001), which had to follow that the size of subsequent dinoflagellate cyst generations became increasingly less. Tertiary cysts are usually smaller than its predecessors from the Mesozoic and reach a size usually between 20 and 75 μm, this applies to most types (Wendler, 2001, 2002; Wendler and Willems, 2002).

According to their life-cycle, the group of calcareous cyst forming dinoflagellates can be subdivided into two groups (Tangen et al., 1982; Fensome et al., 1993; Meier et al., 2007). The first group produces fossilisable calcareous resting cysts during their life cycle (sexual reproduction phase), representing a dormant stage in which metabolism is greatly reduced (Binder and Anderson, 1990), alternating with a non-fossilisable thecate swimming stage (Wall and Dale, 1968; Lewis, 1991), in which they spend most of their life time (Fig. 2).

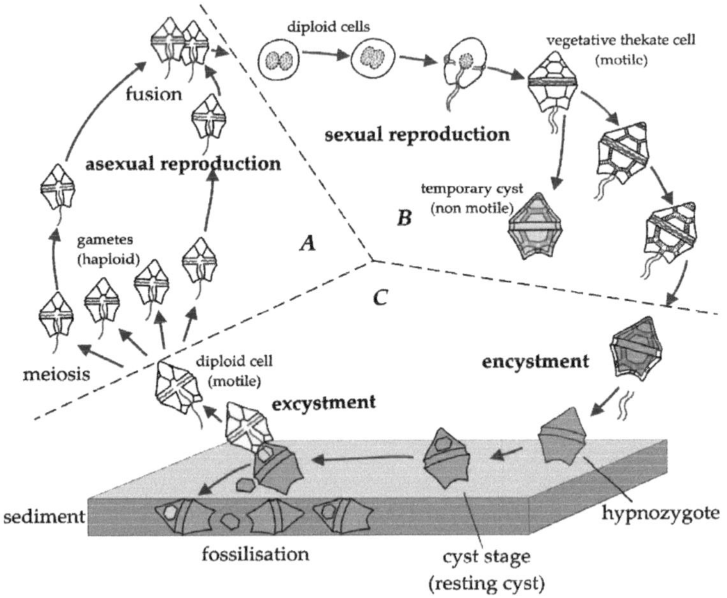

Fig. 2: Simplified dinoflagellate life cycle of resting cyst producing dinoflagellates, involving sexual reproduction. **(A)** Cells are motile and haploid. **(B)** Cells are motile and diploid (nucleus dotted). **(C)** Cells are nonmotile (except for hatched cell) and diploid. Hatched area in discarded cyst at the left represents the archaeopyle (modified after Fensome et al. 1996 and Vink, 2000).

It is suggested that the resting cyst stage is a benthic phase in the life cycle, where dinoflagellates can survive for long periods in the sediment (Anderson et al., 1987; Dale, 1983; Halgraeff and Bolch, 1992). Based on resting cyst studies of stored sediments, Lewis et al. (1999) suggested a minimum of 10 years. Nevertheless, it is questionable whether all species go through a benthic resting cyst stage. It is most likely that some open oceanic species pass this stage in the upper water column (e.g. Harland, 1983; Dale, 1986, 1983, 1992), or they do not form resting cysts at all, such as *Thoracosphaera heimii*.

In the second group (i.e. *T. heimii* and *Leonella granifera*), the fossilisable vegetative non motile cyst (coccoid) stage (asexual reproduction phase) - which is metabolically and/or reproductively active - dominates their life time (Tangen et al., 1982; Inouye and Pinaar, 1983; Fensome et al., 1993; Meier et al., 2007) (Fig. 3). Fossil cysts are generally considered as resting cysts (Wall and Dale, 1968), except those formed by the calcareous species *T. heimii* and *L. granifera*, which produce fossilisable vegetative cysts (Meier et al., 2007). Interestingly, a cyst operculum, which is believed to be an apomorphic feature (see Foissner et al., 2007) and apparently obligatory for the resting cyst phase, has - as far as I know - never been documented for fossil *T. heimii* cysts, although it was observed in culture studies (Tangen et al., 1982). For fossil *L. granifera* cysts, on the other hand, the presence of an operculum is a common character. This suggests - or at least it can not be ruled out completely - that *L. granifera* also passed, or may be still passes through a resting cyst phase during its life cycle. Since in micropalaeontological studies the actual function of the cysts usually can not be determined, all fossil calcareous dinoflagellate taxa are thus generally termed as cysts or calcareous dinoflagellates (dinocysts).

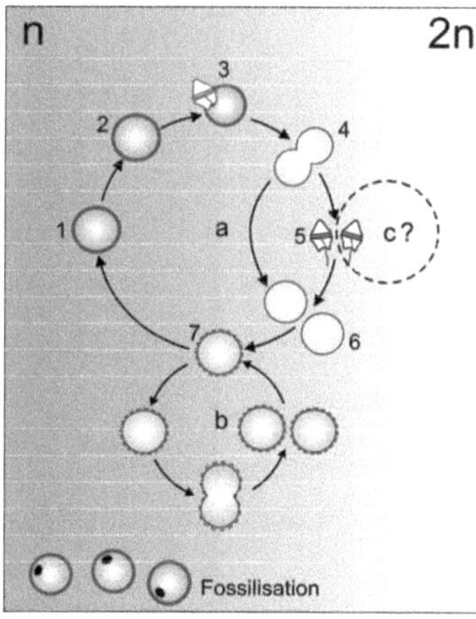

Fig. 3: Simplified life cycle of *L. granifera* and *T. heimii.* **(n)** haploid, **(a, 1-2)** vegetative calcified cysts = dominant life cycle, **(a, 3)** motile cell hatches, **(a, 4)** cell divides, **(a, 5)** forms aplanospores first, **(a, 6)** or dirctly planospores, **(a, 7)** calcification starts, **(b)** mitotic division of weakly calcified cells. Sexual reproduction might occur in a separate life-cycle stage **(2n)** diploid, **(c)** (modified after Meier et al., 2007).

In modern oceans, *T. heimii* is widespread and usually significantly dominates the calcareous dinoflagellate assemblages because of its high reproduction rate of up to one division per day (Brand and Guillard, 1981; Meier et al., 2007), except in the Arabian Sea, where *L. granifera* can dominate (Wendler et al., 2002). Although *T. heimii* already has been documented since the Upper Cretaceous (Hildebrand-Habel and Willems, 2000), it usually was very rare throughout the Tertiary. Referring to our investigated time interval, of the late Miocene and early Pliocene Mediterranean, we can say that *T. heimii* was rather underrepresented, suggesting a much lower reproduction rate than today, or it had a different main life cycle stage, e.g. the non-fossilisable thecate swimming stage. Another explanation for the low occurrence of *T. heimii* during that time is that it lacked the ability to form resting cysts, which is considered to be a key strategy for survival and spread through space and time (Hallegraeff and Bolch, 1992). Nowadays, the survival strategy of *T. heimii* seems to be its high reproduction rate.

Application

Calcareous dinoflagellate cysts are known at least from the early Triassic period (Janofske, 1992) and have left an extensive fossil record since the end of the Triassic (e.g. Fütterer, 1977, 1984; Keupp 1981, 1987, 1991; Fensome et al., 1993, 1999; Kohring, 1993a, 1997; Kienel, 1994; Zügel 1994, Hildebrand-Habel and Willems, 2000; Streng et al. 2004; Bison et al., 2007, 2009). Their highest diversity they reached during the Cretaceous (e.g. Keupp, 1991; Kohring, 1993), followed by a progressive decrease throughout the Tertiary, finally resulting in a relative manageable group of nowadays (Keupp, 1991; Lewis, 1991; Montresor, 1995).

Although calcareous dinoflagellate cysts seem to be a common element in surface sediments around the world oceans (Montresor et al., 1998), in the past major interest was given to organic walled cyst producing dinoflagellates. Calcareous cysts were often overlooked in micropalaeontological studies, either due to there size, which falls between the larger foraminifers and the smaller coccolithophores, or due to the commonly used acid treatment in palynological studies. However, since Deflandre (1947) established the family of Calciodinelloideae, over the last decades renewed interest occurred in calcareous cyst producing dinoflagellates, stimulating the research on their distribution in the surface waters and sediments of the worlds ocean and their relation with the sea surface parameters (e.g. Fütterer, 1977; Lewis, 1991; Montresor et al., 1997; 1994; Höll et al., 1998, 2000; Karwath,

2000; Zonneveld et al., 2000; Vink et al., 2000, 2002; Wendler et al., 2002a, b, c; Vink, 2004; Richter et al., 2007; Elbrächter et al., 2008). As a consequence, knowledge about calcareous dinoflagellates, their ecology, biology and environmental relations remarkably increased over the last years. It was shown that their distribution in the oceans is strongly dependent on the environmental parameters of the surface water (e.g. temperature, salinity, nutrients and water turbulence). Therefore, there fossils reflect the conditions which prevailed during their life time in the surface waters. Apparently, this makes calcareous dinoflagellates very useful for palaeoenvironmental reconstructions. In addition, their high preservation potential and their long stratigraphic range - most of the main Tertiary calcareous dinoflagellates are still extant - confirm their utility as a proxy for long- and short-term palaeoenvironmental studies.

Previous palaeoenvironmental studies

Despite from the high-quality fossil records, so far, mainly inventory taking studies on fossil calcareous dinoflagellates have been carried out, with a main focus on descriptive taxonomy. Most of the studies are from the Mesozoic to Tertiary European and Atlantic realm (e.g., Bolli, 1974; Keupp, 1981, 1987, 1991, 1993, 1995; Keupp and Mutterlose, 1984; Keupp and Versteegh, 1989; Kienel, 1994; Zügel, 1994; Willems, 1996; Hildebrand-Habel and Willems, 1997). These initial studies on calcareous dinoflagellates, which are well documented by numerous scanning electron microscope (SEM) photographs, give us a picture of the amazing variety of calcareous dinoflagellate cyst shapes. A few, mainly long-term evolution studies, related to climatic change, exist from sediments of late Cretaceous to Cenozoic age, taken from the Northern European realm and the Atlantic and Indian Ocean (e.g. Keupp, 2001; Keupp and Kowalski, 1992; Kohring, 1993a; Hildebrand-Habel and Willems, 2000; Hildebrand-Habel and Streng, 2003; Streng et al., 2004). The majority of palaeoenvironmental studies using calcareous dinoflagellates, were performed on Quaternary surface sediments from the South and Equatorial Atlantic Ocean (e.g. Zonneveld et al., 1999, 2000; Esper et al., 2000; Höll et al., 2000; Vink et al., 2001). Other studies focused on the distribution of calcareous dinoflagellates in relation to the environmental parameters of the surface waters (Karwath et al., 2000; Meier and Willems, 2003; Vink, 2004; Richter et al., 2007; Richter, 2009).

From the Mediterranean realm, fossil (pre-Quaternary) palaeoenvironmental studies on calcareous dinoflagellates were so far the subject of only a few and patchy works (e.g. Keupp and Kohring, 1993, 1999; Keupp et al., 1994; Kohring, 1993, 1997). From the Pleistocene, there exist only two studies on calcareous dinoflagellates related to the sapropel

S1 formation (Zonneveld et al., 2001; Meier et al., 2004). Most of the aforementioned fossil studies have in common a relatively low sample density and a mostly semi-quantitative approach, based on a time consuming picking method, which will be discussed later. Our present knowledge of recent to sub-recent Mediterranean calcareous dinoflagellates is mostly based on studies from surface sediments of the Gulf of Naples (e.g. Montresor and Zingone, 1988; Montresor, 1995, Montresor et al., 1998) and on one study (Meier and Willems, 2003), which certainly covers most parts of the Mediterranean Sea. These studies provide an important basis about the distribution and environmental relations of calcareous dinoflagellates in the present Mediterranean Sea. A more detailed analysis of the Mediterranean dinoflagellate association is still desirable though.

In order to increase the existing knowledge about fossil calcareous dinoflagellates from the Mediterranean realm and to prove their applicability for palaeoenvironmental reconstructions, this study aims to present the first detailed analysis of its associations in relation to environmental changes caused by the Messinian salinity crisis (MSC), temporally and spatially.

Classification and identification

Calcareous dinoflagellates (Thoracosphaeraceae, Dinophyceae) are considered to be a monophyletic group of peridinoid taxa that have the potential to produce calcareous cysts during their life cycle (Elbrächter et al., 2008). Already in 1947, Deflandre established the family Calciodinellidae, based on the description of *Calciodinellum operosum* Deflandre, 1947. Later on, Wall et al. (1970) and Keupp (1981) considered the two subfamilies Peridinioideae and Calciodinelloideae as a monophyletic unit (see Keupp, 1984). Subsequently, all fossil and recent calcareous dinoflagellates were summarized in the subfamily Calciodinelloideae Deflandre 1947, regardless of the tabulation pattern (Keupp, 1984). In the following years, several factors such as the later inclusion of non-tabulated forms, the discovery of a coccoid vegetative stage of *T. heimii* (Tangen et al., 1982) and the strong phenotypic variability of the cyst taxa have caused a confusing chaos in the classification of calcareous dinoflagellates (Keupp, 1981).

During the last decades various attempts have been made to unify the systematic concept of dinoflagellates, in general and with respect to calcareous dinoflagellates (e.g. Bujak and Davis, 1983; Fensome et al, 1993; Gottschling et al., 2005; Streng et al., 2006). Two previously independent classification systems have been used for dinoflagellates thitherto, a palaeontological (fossil cyst-based) and neontological (recent motile cell-based) (e.g. Fensome et al., 1993; Elbrächter et al., 2008). Fensome et al. (1993) proposed a single classification system that addresses both, fossil and extant dinoflagellates, following the International Botanical Code of Nomenclature (IBCN). Therein plate tabulation is considered to be one of the most important morphological characteristics at the family, subfamily and genus levels. In particular, pioneering work of Fütterer (1976, 1977) and Keupp (1979b, 1981) have shown that a central feature for the classification and identification of calcareous dinoflagellates is the orientation of the crystallographic c-axis of the wall crystals, at least at the species level (e.g. Keupp, 1981, 1987, 1991; Kohring, 1993; Janofske, 1996; Montresor et al., 1997; Meier et al., 2009) (Fig. 4). This trait is species specific and genetically fixed (Addadi and Weiner, 1989; Addadi et al., 1990; Sikes et al., 1994).

Accordingly, four types of wall structures have been established by Young et al. (1997), based on the concept of Keupp (1981, 1987) and Kohring (1993), i.e. 1. irregularly oblique, 2. regularly radial, 3. regularly tangential, 4. regularly oblique (pithonelloid) (Fig. 4). The pithonelloid wall type (Keupp, 1987; Young et al., 1997) belongs to an extinct group of uncertain dinoflagellate affinity (Fensome et al., 1993; Streng et al., 2004b; Kohring et al., 2005). The orientation of the c-axes of the wall forming crystals does not necessarily correspond with its morphology (Janofske, 1996, Janofske and Karwath in Karwath, 2000). Thus, for the determination of the crystallographic c-axis the light optical analysis has been established (Janofske, 1996).

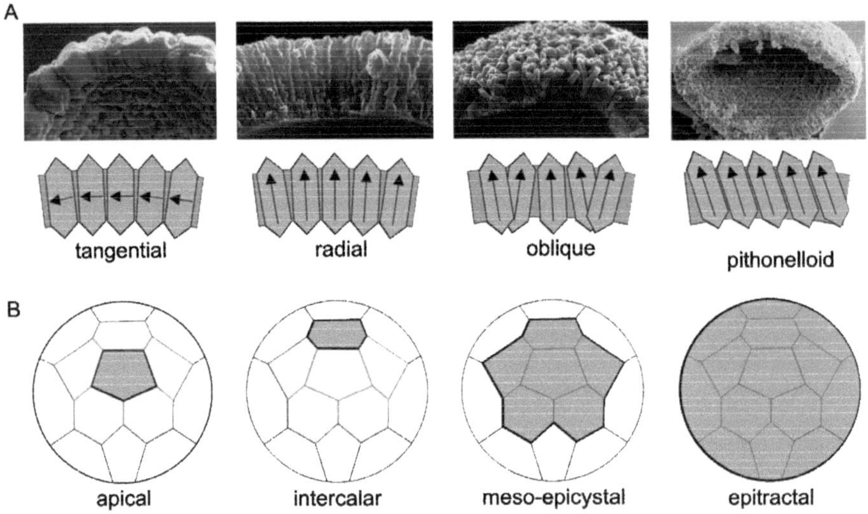

Fig. 4: Taxonomically important morphological characteristics of calcareous dinoflagellate cysts. **A:** Schematic figures and SEM images of the four wall types, showing the crystallographic ultrastructures and orientations of c-axes of the cyst wall (after Kohring et al., 2005). The SEM images of the wall structure show from left to right: *L. urania*, *C. stella*, *P. parva* and *Pithonella cardiiformis*. **B:** Archaeopyle types (after Streng et al., 2004).

Another taxonomically significant feature is the structure of the archaeopyle, (Keupp and Versteegh, 1989; Streng et al., 2002, 2004), which often is the only indicator for the tabulation (Streng et al., 2004). Based on the principle categories of archaeopyle types proposed for organic walled dinoflagellates (e.g. Evitt, 1967; Williams et al., 1978, 2000; Fensome et al., 1996), Streng et al. (2002, 2004b) established a classification model for a variety of archaeopyle types, including a common descriptive terminology supplemented by a few new terms for specific characteristics (Fig. 4).

In recent years, the current morphological and ultrastructural issues are increasingly supplemented by genetic aspects (Gottschling and Plöttner, 2004; Gottschling et al., 2005, 2008). It has been shown that the molecular characterization of dinoflagellates in most cases confirms the external morphology (e.g. Streng et al., 2002, 2004b). The genetic approach in palaeontological studies apparently offers a new concept for testing trait evolution in relation to phylogenetics and environmental change (Dale, 2001). However, as fossil calcareous dinoflagellates typically leave no genetic material behind and the classification of fossil taxa is usually based on morphology, the accuracy of establishing real relationships between extant and fossil taxa is difficult. On the other hand, fossil taxa are commonly interpreted in the light of our knowledge of extant species (Jenner et al., 2008). Concluding, it can be said that in micropalaeontological studies the exact identification (classification) of taxa requires a combined method that considers morphological (e.g. outer shape, wall structure, form of the archaeopyle, tabulation) and ultrastructural (e.g. orientation of the c-axis) features. Within this study the genetic aspects have no relevance (yet).

Methods

Within micropalaeontological studies no single method of study can give us all the information we require for exact identification and interpretation. And two different methods will never give us exactly the same set of information and in most cases there will be significant differences in the results obtained by different methods. Whichever method we choose ultimately depends on many factors and requires a thorough prior review.

A new method for the quantification and identification of fossil calcareous dinoflagellates using the SEM

There are two basic analysing techniques for the quantification and identification of calcareous dinoflagellate cysts. The first uses the polarized light microscope (PLM) and the second the scanning electron microscope (SEM). The orientation of the crystallographic c-axes can only be examined by the PLM, whereas the outer shape and details of the wall structure of the cyst and its modification through diagenetic processes is only visible by SEM analyses. In the past, fossil dinoflagellate analysis was mainly carried out by semi-quantitative SEM investigations (e.g. Keupp, 1981; 1982; Kienel, 1984; Kohring, 1993, 1997; Keupp and Kohring, 1999; Zügel, 1994; Hildebrand-Habel and Willems, 1997, 1999, 2000; Hildebrand-Habel and Streng, 2003). It based on a combination of picking single specimen under the reflected light microscope (RLM) and it's afterward counting (semi-quantitative analysis) under the SEM (see Hildebrand-Habel and Willems, 1999). This method is very time consuming and therefore not competitive in micropalaeontological studies today.

For recent and sub-recent (Quaternary) material, the PLM technique (Janofske and Keupp, 1992; Janofske, 1996; Young et al., 1997) has been widely used during the last years to quantify total calcareous dinoflagellate cyst composition from water samples and sediment surface samples of relative loose material with low compaction and diagenetic overprint (e.g. Wendler et al., 2002a, b, c; Meier and Willems, 2003; Vink, 2004; Richter et al., 2007). It was a first step in using calcareous dinoflagellates for palaeoenvironmental studies (Zonneveld et al., 2005). Currently, the PLM technique remains as the best instrument for the determination of relative loose and unconsolidated recent to sub-recent material since it is inexpensive, readily available and enables fast identification for the main (~12 species) recent calcareous dinoflagellate species (see Zonneveld et al., 2005).

For older fossil material, the analysis with the PLM holds some problems. For example, with the PLM most common species are easily recognised only, if they are not modified by secondary crystal growth. For fossil material, suffering a diagenetic overprint, safe identification on species level is difficult, if not impossible for some specimen without calibration under the SEM. The following problems exist:

- Outer shape, details of the surface and wall structures and its modifications through diagenetic processes can not be detected with the light microscope.
- Morphological modifications (e.g. increased wall thickness) lead to deviations from the mean phenotype (i.e. species specific colour distribution), as light can't pass through the cyst.
- Many fossil species have not yet been detected by light optical analyses. For these species the morphological features need to be classified using the SEM.
- Shape and size of the cyst vary from species to species. In addition, intraspecific morphological cyst variations (ecophenotypism) are a common feature and can be interpreted as a response to changing environmental parameters. Therefore, the detailed detection of the cyst morphology can provide complementary information for palaeoenvironmental reconstructions.

In order to find out the most suitable method for our fossil material, we did some pilot studies, comparing both techniques. After a thorough check of the results, the SEM analysis turned out to be more suitable for our study. It has been shown that:

- The time needed to analyse a sample with the SEM and PLM is comparable.
- SEM-analysis allows the identification of a variety of key characteristics (e.g. tabulation, form of the archaeopyle), including outer and inner cyst morphology, structure of the wall surface and of single crystals and its modifications by secondary crystal growth.
- The ability to detect species diversity was much higher with the SEM, than with the PLM.
- Species composition in samples with low diversity was comparable, but much more difficult to detect with the PLM and therefore needed to be counterchecked with the SEM.
- Species with similar ultrastructural and morphological characteristics (e.g. *C. albatrosianum*/*C. levantinum*) were difficult to differentiate, if modified by secondary crystal growth.
- Double checking is advisable for both methods in order to differentiate intraspecific from interspecific morphological variations in order to detect cryptic species.

Within this study scanning electron microscopy was applied for quantification of the dinoflagellate cyst assemblages and as well to obtain reliable species identification (Bison et al., 2007, 2009). Light optical analysis was additionally used to confirm and clarify the species identification in cases of uncertainties (see Bison et al., 2007, 2009). The application of the new method requires different preparation techniques in parallel, but only a one-step sample treatment procedure, which follows the conventional method (e.g. Vink et al., 2000), usually used for Quaternary micropalaeontological studies using the PLM. For selective picking of single specimen, the RLM was used.

In summary, for this study for the first time we primary used the SEM for the quantification and identification of fossil calcareous dinoflagellates and the PLM only occasionally for verification. With this method improvement for the analysis of fossil material, the use of calcareous dinoflagellates for palaeoenvironmental reconstructions becomes more attractive and competitive.

Material in general

The samples of this work derive from various land sections throughout the Mediterranean realm (e.g. NW Morocco (Bou Regreg area); Spain (Sorbas Basin); Sicily (Caltanissetta Basin); N Italy (Monte del Casino section); Cyprus (Pissouri Basin)). They were collected from colleagues of different institutions during different field surveys by (Krijgsman et al., 1995; Hilgen and Krijgsman, 1999; Rouchy et al., 2001). The material from Morocco and Spain we collected during a common fieldwork with colleagues from Utrecht. The sediments range from indurated carbonates to less indurated marls and sapropels. Altogether about 170 samples have been analysed on their calcareous dinoflagellate content. The results of 118 samples are either already published (paper 1, 2) or to be submitted for publication (paper 3). The publication of the other results is still pending and still requires additional counting (see Chapter 5). Images are recorded in the University of Bremen.

Sample preparation procedure and calculation

The sample treatment and procedure is presented here as a laboratory processing protocol. A short description of the method is also given in chapter 2 and 3. The sample processing (treatment and sieving) and the laboratory preparation for the PLM analysis follows the method described by Vink et al. (2000). The sample processing and preparation of the residue sub-sample of the 20 – 75 μm fraction for the analysis of selectively picked cysts under the RLM follows the laboratory processing described by Hildebrand-Habel and Willems (1997, 1999; modified by Bison, 2001).

Part A: Treatment and sieving

1. Dry a sufficient amount of the sediment material in the oven (~75°C) for ~24 hours (use a plastic beaker with lid).
2. Weigh exactly 0,5g of the dried sediment (use the 100 ml plastic beaker) on a high precision balance. If necessary break the sediment into fragments, use a vice (carefully!). Take care about contamination. After every weighing clean working place and implements (e.g. with air pressure and paper cloths); use new paper cloths for every sample.
3. Disaggregation: a) For unconsolidated material: dissolve the sample (0.5 g) in 100 ml of a 6 % Soda solution ($Na_2CO_3x10H_2O$). b) For samples with elevated organic matter contents (e.g. sapropels): add 10 – 15% H_2O_2 solution for organic matter oxidation. To speed up the oxidation process, briefly heat the suspension to 50°C (take care that the solution does not boil over). Afterwards it can be filled up with water. c) For strongly consolidated material: treat the suspension with repeated freezing and thawing in ~6 % Soda solution (e.g. Zügel, 1994; Willems, 1996; Hildebrand-Habel and Willems, 1997).
4. Briefly (few seconds) tread the suspension with ultrasound (Sonorex RK 100, 35 kHz frequency) for homogenising. Repeat if necessary.
5. We primarily keep the fraction 20 – 75 µm. The fractions 5 - 20 µm (following Vink et al., 2000) and > 75 µm were tested only randomly for their calcareous dinoflagellate content. Wash the suspension thoroughly with tab water through a 75 µm steel sieve into a 3000ml glass beaker (the size is irrelevant but it is suitable for thorough sieving). Take care that nothing gets lost. Wash the suspension (0 – 75 µm) collected in the beaker glass (without settling down) through a 20 µm steel sieve. The fraction >75 µm can be transferred into a porcelain bowl, dried in the oven and checked for calcareous dinoflagellates using the RLM.
6. Transfer the sediment of the 20 – 75 µm fraction, remaining on the sieve, carefully into a scaled 20 ml test tube with lid. Use a cone. Add a few drops of ammonia and ethanol to avoid contamination with fungi and to reduce surface tension (Vink et al., 2000). Let it settle down for a few hours.

7. Reduce the water column in the 20 ml test tube (20 – 75 µm fraction) with a glass pipette to a sample volume of 12 or 15 ml, depending on the amount of sediment.

Part B: SEM-stub preparation

1. For the quantitative analyses, prepare a storage box for the safe keeping of the SEM-stubs (Careful storage and labelling is required).
2. Label every SEM-stub at the bottom with a permanent marker.
3. Apply a double-sided adhesive lead-tab onto the SEM-stub.
4. Remove the protection film from the lead-tub and apply a round cover glass (13 mm Ø).
5. Place the SEM-stubs on a conductive multi-stub holder.
6. Remove 50 µl (100µl) from exactly 15 ml (12ml) - at a depth of 1 cm below water surface - from the thoroughly homogenised 20 – 75 µm fraction of the test tube. Use an Eppendorf pipette. The remaining sub-sample can be placed in porcelain bowls and oven-dried for later selective picking.
7. Transfer the sub-sample on the cover glass of the labelled SEM-stub placed on the stub-holder.
8. Check the distribution of the material under the RLM.
9. For reduction of the surface tension and for a better dispersion of the 50 µl sub-sample add (very carefully!) - if necessary - one small drop of ethanol using a pipette.
10. For drying place the multi-stub-holder with the sub-samples shortly on a warming plate (~100°C).
11. Sputter the sub-samples with gold prior to SEM examination. The sample is now ready for counting using the SEM. SEM conditions were 15 – 20 kV, working distance 12 - 15 mm.

Part C: PLM-stub preparation

1. For the PLM analyses prepare a slide-tray for the safe handling of the slides during the sample processing.
2. Label every slide on the sides and keep the middle free.
3. Place the slides on the slide-tray.

4. Remove 100 µl (50µl) from exactly 15 ml (12ml) - at a depth of 1 cm below water surface - from the thoroughly homogenised 20 – 75 µm fraction of the test tube. Use an Eppendorf pipette. Sub-samples from the small fraction of 5 – 20 µm can be placed on a second slide.
5. For drying place the slide-tray with the sub-samples in the oven (~50°C).
6. Embed the dried sub-sample on the slide in Spurr's resin (Spurr, 1969) and directly put a cover glass on it.
7. Place the slide-tray with the slides (sub-samples) overnight for hardening in an oven with flue. The sample is now ready for counting using the PLM.

Part D: RLM preparation for selective SEM analysis

1. For the RLM analysis provide a holder for the SEM-stub and Fema-cell (Bison, 2001), a storage box for the safe keeping of the SEM-stubs and a petri-dish.
2. Label the microslides and SEM-stubs.
3. Apply a double-sided adhesive lead-tape onto the SEM-stub.
4. Fix the SEM-stub (use the clamp forceps) on the holder (support table), use a screwdriver; remove the protection film from the lead-tape and carve a grid in the lead-tape.
5. Place the picking tray and the microslide below the RLM.
6. Scatter the loose material on the picking tray (use a brush).
7. Pick the specimen with the eyelash and transfer them into the microslide.
8. Place the microslide on the holder with the equipped SEM-stub.
9. Place the fully equipped holder below the RLM and adjust the SEM-stub holder so that both carrier units (microslide and SEM-stub) are in the right focus (fix the holding device with the screw).
10. Transfer the single specimen from the microslide to the SEM-stub.
11. After sputter coated with gold the picked specimen can be analysed and photographed under the SEM.

Abundance calculation

Calcareous dinoflagellate cysts were studied and photographed using a scanning electron microscope (SEM) in combination with a polarized light microscope (PLM) supplemented by a reflected light microscope (RLM) for selected picking. Cysts were counted under the SEM and the number of counts was used to calculate the number of cysts per gram sediment in the 20 - 75 µm fraction (see chapter 2 and 3). In general we prepared four SEM-stubs for each sample. For most samples multiple counts were made to reach a statistical relevant outcome for the quantitative calculation. However, for some samples we could not meet this intent, as they did not contain enough cysts. In parallel we usually prepared two slides for the 20 – 75 µm fraction and for randomly selected samples one slide for the 5 – 20 µm fraction for the verification under the PLM. Relative abundance has been calculated as the abundance of a species (by any measure), divided by the total abundance of all species combined. Diversity has been calculated after the Shannon-Weaver Diversity Index (see chapter 2).

Repository

The studied material is deposited in the collection of the Division of Historical Geology and Palaeontology, Department of Geosciences, University of Bremen, Germany.

The Mediterranean Sea

„Dredging was our bête noire. The romance of deep-sea dredging or trawling in the Challenger, when repeated several hundred times, was regarded from two points of view; the one was the naval officer's who had to stand for 10 or 12 hours at a stretch, carrying on the work... the other was the naturalist's ... to whom some new worm, coral, or echinoderm is a joy forever, who retires to a comfortable cabin to describe with enthusiasm this new animal, which we, without much enthusiasm, and with much weariness of spirit, to the rumbling tune of the donkey engine only, had dragged up for him from the bottom of the sea." (From a naval officer's diary)

Geologic history

The present heterogeneous geological structure of the Mediterranean region is the result of a highly complex tectonic evolution and interaction between Eurasia and Gondwana, with successive episodes of continental break-up, ocean development, subduction, continental collision and orogeny, spanning some 250 million years, from the late Permian to the Quaternary (e.g. Morris and Tarling, 1996; Cavazza et al., 2004). The tectonic system is dominated by connected fold- and thrust-belts, associated with the development of different oceanic basins of variable size and age (Stampfli and Borel, 2004). The opening and subsequent consumption of two major oceanic basins – the Palaeotethys (mostly Palaeozoic) and the Neotethys (late Paelaeozoic–Mesozoic) – was accompanied by the establishment of individual smaller oceanic basins within an overall regime of prolonged interaction between the Eurasian and African-Arabian plates finally resulting in its present day geological configuration (Robertson, 1998; Cavazza et al., 2004). Therefore, the Mediterranean Sea can be seen as a remnant of the Neotethys Ocean (Robertson, 1998).

During the Cretaceous it formed a continuous passageway between the Indo Pacific and Atlantic Ocean. This seaway separated Africa from Eurasia throughout the Mesozoic and continued as a barrier until the early to middle Miocene, when the eastern connection to the Indo-Pacific Ocean closed, forming the Mediterranean Sea (e.g. Adams et al., 1983; Vergnaud-Grazzini, 1983; Rögl, 1999). Since the eastern closure of the Tethys, the Mediterranean has continued to narrow as the result of the still progressing northward migration of the African plate, accompanied by the subduction beneath Italy and the Greek Islands and the collision of Arabia and Iran with Asia. The present Atlantic-Mediterranean connection through the Strait of Gibraltar originated in the early Pliocene. Previously the Atlantic inflow took place via two gateways, the Betic Corridor (southern Spain) and the Rifian Corridor (northern Morocco). The connection via the Betic Corridor closed during the Tortonian and resulted in a so called Tortonian salinity crisis (TSC), whereas the closure of the Rifian Corridor induced the Messinian salinity crisis (MSC). Today, the Mediterranean area provides a present-day geodynamic analogue for the final stage of a continent-continent collisional orogeny, eventually ending with the final elimination of the Mediterranean Sea (Cavazza et al., 2004).

Modern Mediterranean Sea

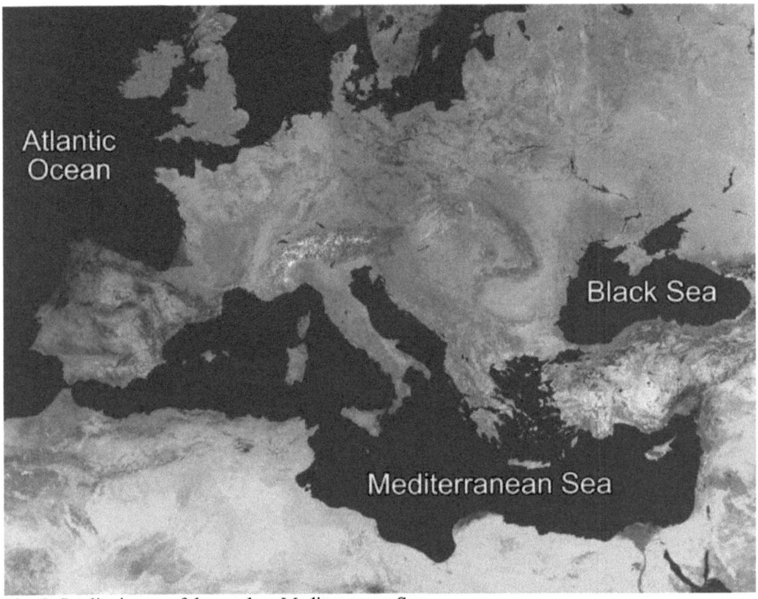

Fig. 5: Satelite image of the modern Mediterranean Sea.

Nowadays, the Mediterranean Sea is a deep (average ~1500 m) large almost enclosed basin - highly complex in many purposes, e.g. in its geological structure, geography, climate and hydrology - bounded to the north by Europe, to the south by Africa and to the east by Asia (Fig. 5). Its only natural connection to the open Ocean is through the Strait of Gibraltar – a shallow sill of about 320 m depth - enabling the water exchange with the Atlantic Ocean. An about 400 m deep sill at the Strait of Sicily divides the Mediterranean Sea into the western and eastern basin. Via the Suez Canal it is connected to the Red Sea, and via the Bosporus and the Dardanelles Straits to the Black Sea. These straits and passages are key regions since they control the mass transport triggered by the thermohaline circulation (Beranger et al., 2005). The main characteristics of the Mediterranean Sea are its anti-estuarine circulation, the high oligotrophy, high salinity and strong seasonality.

The anti-estuarine thermohaline circulation is established by the inflow of relative cold and nutrient depleted normal saline Atlantic Ocean waters (AW), entering the Mediterranean Sea through the Strait of Gibraltar as a surface water flow (Beranger et al., 2005 and citations herein). It spreads out eastwards throughout the entire basin. During this eastward flow, temperature and salinity increase whereas nutrients decrease, leading to the characteristic east-west gradient. After its arrival in the eastern Mediterranean basin, one portion of the Atlantic water flows to the northern coast (Adriatic Sea), where it cools down (~13°C) and sinks during winter. Another portion of the remaining Atlantic water continues to flow towards Cyprus, where it also sinks during winter to a depth ranging from 200 to 600 meters, forming the Mediterranean Intermediate Water (MIW). This relatively nutrient enriched water mass then flows back into the Atlantic Ocean through the Strait of Gibraltar. As a result of this circulation, on average the Mediterranean Sea established its typical highly oligotrophic conditions with a negative gradient to the east.

High seasonality in the surface water temperature (SST) is another characteristic of the modern Mediterranean Sea. During the winter season SST ranges from 11 – 13 C° and from 25 – 30 C° during summer (Beranger et al., 2005 and citation herein). In contrast, the deep-water temperature remains relatively constant throughout the year, with an average of ~13 C°. During the summer, the thermocline divides the variable surface waters from the more stable deep-waters and thus prevents downwelling. The high seasonality is caused by its geographical position, which brings it under the descending branch of the Hadley circulation (very dry air) in summer, while the Westerlies (wet wind belt) prevail during the winter season (Bolle, 2003). On a global scale, the Hadley circulation is coupled to the zonal tropical circulation (Southern Oscillation), which is interconnected with the Indian Monsoon, which in turn influences the eastern Mediterranean area (Bolle, 2003). The complex structures of the surrounding land masses and the extreme climate gradient across the Mediterranean Sea makes it difficult to generalize and model the Mediterranean climate (Bolle, 2003). However, on average it is characterized by mild to cool rainy winters and warm to hot dry summers, driven by the seasonal alternation between cyclonic storms in winter and subtropical high pressure cells in summer. Besides seasonal surface water cooling, high salinity is the other main factor driving the anti-estuarine circulation. In the Mediterranean evaporation greatly exceeds precipitation and river runoff (e.g. Pinet, 1998; Bethoux and Gentili; 1999). Thus, salinity reaches up to high values of 39 psu in the eastern basin. Although rainfall and rivers add 1830 km^3 to the Mediterranean every year, it loses 4690 km^3 every year due to evaporation. The water-loss has to be balanced by the Atlantic water inflow, otherwise the

Mediterranean Sea will desiccate within some thousand years, leaving huge amounts of evaporites (Meijer, 2006). Exactly this has happened about 6 Ma ago, during the so called Messinian salinity crisis (MSC).

The Messinian salinity crisis (MSC)

"Strange and beautiful things were brought to us from time to time, which seemed to give us a glimpse of the edge of some unfamiliar world." (C. Wyville Thomson, The Challenger Expedition, 1876)

The Messinian salinity crisis (MSC) (Selli, 1960) was a geological event at the end of the Miocene, between 5.96 and 5.33 Ma, in which the Mediterranean - Atlantic connection progressively got restricted and interrupted, triggered by tectonic uplift processes in the Gibraltar area (Van Assen et al., 2006). During this time, the Mediterranean realm underwent dramatic environmental changes, resulting from a complex interplay between tectonic and climatic factors (see CIESM, 2008). Huge amounts of evaporites, up to 1500 m thickness (Montadert et al., 1970) and of estimated one million km^3 volume in total (Ryan, 2009), were deposited on the Mediterranean seafloor within a relatively short time period (Meijer and Krijgsman, 2005), apparently witnessing the deep desiccation of the Mediterranean basin (Fig. 6). Almost 40 years after formulating the so called "deep-basin desiccation model" (Hsü et al., 1973), numerous scenarios of the onset, timing and development of the MSC have been controversially discussed (e.g. Hardie and Lowenstein, 2004; Rouchy and Caruso, 2006; CIESM, 2008; Roveri et al., 2008a; Ryan, 2009).

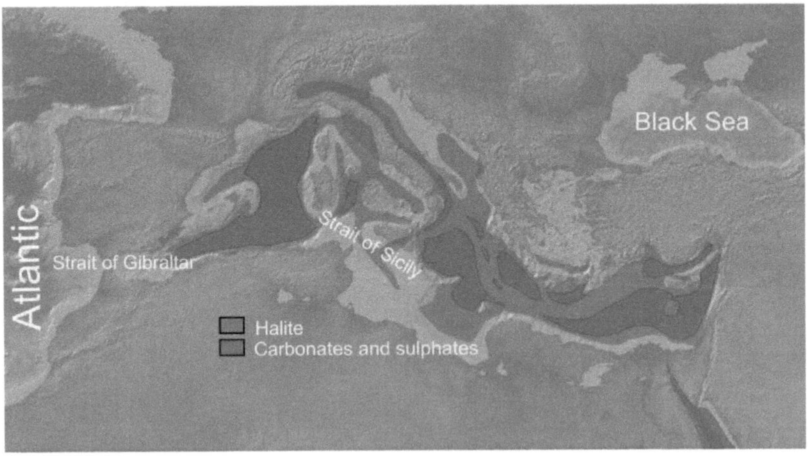

Fig. 6: Evaporite distribution in the Mediterranean area (from Ryan, 2009).

Many aspects are by far not fully understood yet, for example the influence of climatological and oceanographic parameters, the physical geography of the Mediterranean basins and the hydrological structure of the water column during that time (e.g. Blanc, 2000; Roveri et al., 2004). In other words, in the course of the MSC the complex interference of tectonics, eustasy and climate led to the formation of varying lithological successions whose interpretation remains still controversial (e.g. Rouchy, 1982; Clauzon et al., 1996; Orszag-Sperber et al., 2000; Rouchy et al., 2001; Sierro et al., 2001; Fortuin and Krijgsma n, 2003; Roveri et al., 2003; Aguirre and Sánchez-Almazo, 2004; Bertini et al., 2006; CIESM, 2008; Ryan, 2009). Several possible courses have been considered to explain the different evolutionary stages of the Messinian crisis and its immense residues, such as the huge evaporite deposits, which occur both, submerged in the deep modern Mediterranean basins (Hsü *et al.*, 1973, 1977; McKenzie, 1999) and uplifted in sedimentary successions on land (e.g. Roveri et al., 2003, 2007, 2008) and the coeval fluvial cutting of subaerial canyons during the desiccation phase (e.g. Chumakov, 1973; Cita and Ryan, 1978; Barber, 1981). One example for an enormous outcrop is the ''Vena del Gesso'', an about 15 km long gypsum belt in the Romagna Apennines (Roveri et al., 2003). Prominent examples of intense fluvial erosions in the south eastern Mediterranean region across North Africa are the Nile (Chumakov, 1973; Barber, 1980) and the Sahabi canyon (Barr and Walker, 1973; Griffin, 1999, 2002).

Although the fact that there was an immense evaporation/precipitation event could not be dismissed and the desiccation model gained widespread acceptance, deep ambiguity was and still persists about the processes and depths at which the evaporites were deposited (e.g. Roveri et al., 2003; Hardie and Lowenstein, 2004; CIESM, 2008; Ryan, 2009). According to the highly complex and heterogeneous structure of the Mediterranean basin, the deposition took place in a variety of tectonic settings within a relatively short time period (Bertoni and Cartwright, 2006). Different models have been developed during the last decades, with three of these have emerged as the most popular (Krijgsman and Meijer, 2008) (Fig. 7). The first one is the initial deep basin - shallow water model, with a major draw down of the Sea level (>1000 m) (e.g. Hsü, 1972a, b, 1973, a, b, 1978b; Cita et al., 1978; Rouchy and Saint-Martin, 1992; Rouchy and Caruso, 2006), the second assumes a shallow basin with shallow waters, by a minor (~150 m) sea level lowering, with evaporite deposition restricted to marginal basins (Clauzon et al., 1996; Roveri and Manzi, 2006) and the third, most recent one, is the deep basin – deep water model, with hardly any sea level lowering during most of the time (Van Couvering et al.,1976; Blanc, 2000, 2006; Lu and Meijer, 2006; Hardie and Lowenstein,

2004; Meijer and Krijgsman, 2005; Krijgsman, 2008; Krijgsman and Meijer, 2008; Ryan, 2009). The latter model assumes that the salinity increase is independent from the water depth and not obligatory for the precipitation of evaporites and that an increase in concentration up to periodical saturation is the driving factor. This does not exclude the formation of evaporites, even in shallow settings, as postulated by e.g. Orszag-Sperber et al. (2009) for the successions on Cyprus (Polemi, Pissouri, Tochni Basins), but contradicts the generalization of the true onset and timing of the evaporite deposition in favor of a more sophisticated view of the different basins, considering local features and basin dynamics.

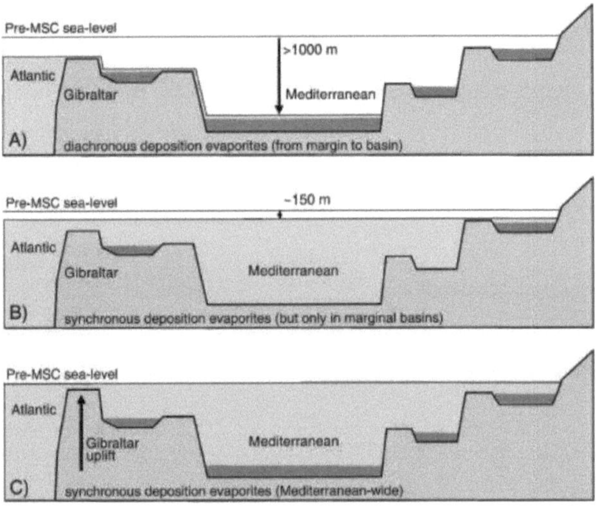

Fig. 7: Different scenarios for the depositional environment of the so-called "Lower Evaporites" of the Mediterranean Messinian Salinity Crisis. Hypotheses range from: **A)** very shallow water and major draw down (N1000m) of the Messinian sea level (Hsü et al., 1973; Rouchy and Caruso, 2006), via **B)** shallow water achieved by a minor (~150m) sea level lowering – with evaporite deposition only in marginal basins – (Clauzon et al., 1996; Roveri and Manzi, 2006), to **C)** hardly any sea level lowering, but uplift in the Gibraltar region, restricting exchange and a the Mediterranean water level similar to the Atlantic level (Krijgsman et al., 1999b; Lu, 2006; from Krijgsman and Meijer, 2008).

The idea that the Mediterranean Sea was strongly restricted during the MSC (5.96 – 5.33 Ma) and at least partly dried out, is nowadays widely accepted, but complete desiccation of the deep basins, as originally suggested (e.g. Hsü et al., 1973), has to be rejected (Roveri et al., 2001; Hardie and Lowenstein, 2004; Manzi et al., 2004; Matano et al., 2005; Ryan, 2009). Most comprehensive and extensive desiccation is estimated for the time interval between 5.59 and 5.50 Ma forming the "Upper Evaporites" (Clauzon et al., 1996; Krijgsman, 1999b, 2008). Krijgsman and Meijer (2008) note that the upper evaporitic phase can be linked to two peak glacials (TG12, TG14) of the Messinian glacial interval (~5.59 – 5.2 Ma) (Van der Laan et al., 2005).

While there is disagreement on all fronts, in particular regarding the different basin models, the most general consensus exists about the biostratigraphic definition of the Messinian stage (Colalongo et al., 1979), the chronology and initial cause of the MSC (e.g. Benson et al., 1995; Hilgen et al., 1995; Krijgsman et al., 1995, 1999, 2010; Hodel et al., 2001, Ryan, 2009). Therefore, the base of the MSC is defined with the onset of the main evaporitic phase which is astronomically dated at 5.96 Ma (Krijgsman et al., 1999a; Hilgen et al., 2000) and its termination is marked by the establishment of fully marine conditions in the Mediterranean, coincident with the Miocene – Pliocene boundary, which is astronomically dated at 5.33 Ma (Lourens et al., 1996).

A key role for the MSC plays the area where nowadays the Strait of Gibraltar is located. Here, the inflow of Atlantic water maintains the present sea level of the Mediterranean Sea. In the past, i.e. during the late Miocene, the Atlantic – Mediterranean water exchange took place via two main gateways, the Rifian (NW Morocco) and the Betian Corridor (SE Spain), before the establishment of the modern connection through the Strait of Gibraltar at the end of the MSC (5.33 Ma) (Benson et al., 1991; Krijgsman et al., 1999). The MSC was initiated by the progressive restriction off these corridors (Benson et al., 1991; Gautier et al., 1994; Krijgsman et al., 1999b). It probably already started during the late Tortonian/early Messinian, i.e. long before the onset of the actual MSC (5.96 Ma) (e.g. Kouwenhoven et al., 1999, 2003, 2006), but most likely culminating just before the beginning of the evaporite deposition (5.96 Ma) (Krijgsman et al., 1999c; Barbieri and Ori, 2000; Van Assen et al., 2006).

The evaporite formation probably occurred in two phases, resulting in the "Lower and Upper Evaporites" (Decima and Wezel, 1973). During the first (regressive) phase (5.96 – 5.59 Ma), widespread gypsum precipitation took place, whereas in the second (transgressive) phase (5.6 – 5.5 Ma) massive salts were deposited. The two evaporate phases are interrupted by the so called Messinian Erosional Surface (MES), commonly interpreted as the product of subaerial erosion, driven by a substantial sea level fall during the climax of the MSC (Rizzini et al., 1978; Ryan and Cita, 1978; Barber, 1981; Stampfli and Höcker, 1989; Field and Gardner, 1991). The MES is visible both on seismic reflection profiles and on outcrops of several Mediterranean sequences (Ryan and Cita, 1978). Eventually, the evaporative phase is followed by a brackish water environment of the so called Lago-Mare period (e.g. Decima andWezel, 1973; Orszag- Sperber et al., 2000; Rouchy et al., 2001, 2003; Orszag-Sperber, 2006; Rouchy and Caruso, 2006; Hilgen et al., 2007; Krijgsman and Meijer, 2008). The MSC finished at 5.33 Ma with the opening of the Gibraltar Strait and the establishment of marine conditions (Krijgsman et al., 1999).

There is agreement that the reestablishment of the Mediterranean-Atlantic connection was primarily triggered by tectonic subsidence at the Gibraltar sill, likely in combination with erosion and sea-level rise (Duggen et al., 2003; Van der Laan et al., 2006; Loget and Van Den Driessche, 2006). Details of the magnitude and extent of this refilling process however, is still under discussion (e.g. Gautier et al., 1994; Krijgsman et al., 1996; Van Dijk et al., 1998).

A further uncertainty factor refers to the climatic conditions persisting before, during and after the MSC in the Mediterranean region and how it contributed to the MSC. For long time the MSC was associated with an overall warm and dry climate with no significant changes before, during and after the MSC (Suc and Bessais, 1990). Recent researches indicate either a warm and humid or a cool and dry climate (Bertini, 2006). Based on pollen data, Bertini (2006) suggests that humid conditions prevailed during the Messinian period however, interfered by wet/dry fluctuations both, in evaporitic and post-evaporitic deposits induced by astronomical precession. Based on a continuous pollen record from the Garraf region (E Spain), Suc and Cravatte (1982) estimated the Early Pliocene (5.33 Ma) climate as warm and humid with an upward tendency toward progressively cooler and drier conditions.

Jimenez-Moreno et al. (2009) imply a tendency towards cooler conditions and a clear latitudinal gradient since the late Mid – Miocene (Serravallian) associated with the beginning of the glacial – interglacial fluctuations. However, for the Mid – Miocene to the Mid – Pliocene they estimated higher mean annual temperatures along the north – south gradient associated with higher precipitation in the north but comparable precipitation to today in the south with a more pronounced gradient compared to today (Jimenez-Moreno et al., 2009). This agrees with the results of Favre et al. (2007), who applied different models in combination with pollen data, estimating temperatures for the Messinian which are on average 2 C° higher than today and thus proposing subtropical to warm - temperate vegetation types in the northern Mediterranean region. This is in agreement with the high contrast between the northern (forest vegetation) and southern (open vegetation) Mediterranean region today.

As for the evaporites, climate models cannot be applied in a generalizing way in the highly diverse Mediterranean region, but always have to take into account local variability. Questions remain also regarding the palaeoenvironmental and palaeohydrographical conditions during the time preceding and following the MSC (latest Miocene and earliest Pliocene). The MSC has left many traces throughout the entire Mediterranean realm. The most suitable and interesting areas to study the environmental impact of the MSC are the following localities: Atlantic Morocco (Bou Regreg section), southern Spain (Sorbas Basin), Sicily (Caltanissetta Basin), Italy (Northern Apennine), Crete and Cyprus (Pissouri Basin). Common characteristic for all these regions is the good accessibility to the Messinian succession, covering in its entirety a complete record of the MSC. These basins, with some exceptions, are well investigated.

This thesis concentrates on the two well exposed Neogene land sections from the Pissouri Basin on Cyprus (eastern Mediterranean) and the Caltanissetta Basin on Sicily (central Mediterranean). Sicily probably represents one of the most suitable areas for studying the MSC on uplifted land sections throughout the Mediterranean realm. All together it provides a complete sedimentary record of the MSC, plus the records preceding and following this event. The sections from the Caltanissetta Basin represent a deep water setting (~1200 m) (Rouchy and Caruso, 2006). Contrary to this, the deposits from the Pissouri Basin represent an intermediate water depth of ~300 - 500 m (Kouwenhoven et al., 2006), also covering the whole history related to the MSC. The sections from Morocco, Spain and Italy will be subject of Chapter 5, where first results will be presented briefly.

Study area

Central Mediterranean basin

After the main phase of the late Mesozoic and Tertiary collision between the African and Eurasian plates, alternation of compressional and extensional tectonics started between the two plate margins (Robertson and Grasso, 1995; Catalano et al., 1996). During the Late Miocene the Mediterranean was segmented into several interconnected basins. The two main basins, i.e. the western and eastern basin, are present since the Early Miocene (~17 Ma ago), separated by the Sicily sill, which probably was deeper during the Messinian than today (Jolivet et al., 2006). In the area, of what is now central southern Sicily, large successor foreland basins (e.g. Caltanissetta Basin) developed in front of the advancing thrust sheets (Maghrebian units) at the end of the Early Miocene (Burdigalian) (Robertson and Grasso, 2005). During the Late Miocene (Serravallian–Tortonian), renewed thrusting led to deformation of the earlier foreland basins and its subsequent southward shift to its present position (Grasso and Butler, 1993). The Messinian configuration of the main basins probably was quite similar to the present, with an always deeper eastern basin. One exception is the southern Tyrrhenian Sea, whose main opening occurred after the Messinian, during the late Pliocene and Quaternary (Blanc, 2000; Govers et al., 2009; Accaino et al., 2010).

Sicily/Caltanissetta Basin

Sicily represents a fraction of the Alpine Collisional Belt (Appenine - Maghrebian thrust belt), which developed between the African and Eurasian plate boundary as a result of several tectonic events, running through the African Maghrebides, Sicily, Calabria, the Appenines and Alps (Catalano et al., 1996; Accaino et al., 2010). On Sicily, one of the largest sedimentary foreland basins of the thrust belt is located, the so called Caltanissetta Basin, which combines a series of thrust related synclines (Butler et al., 1999). This basin covers a large part of central Sicily, extending southwards to the Sicily Channel (see Chapter 4, Fig. 1). It was wedged in a convergent tectonic setting of the Hyblean (Iblean) Plateau to the south (a shallow carbonate platform attached to Africa in the Late Jurassic – Early Cretaceous) and the northern Sicilian Mountain chain to the north (a thickened crustal part of northern Sicily), belonging to the Apenninic - Maghrebian units (Cubito et al., 2005; Catalano et al., 2006; Accaino et al., 2010). It corresponds to the main deposit area of the Maghrebian - Sicilian foreland basins whose evolution was strictly related to that of the Apennine foredeep basins. On Sicily, thrusting and deformation climaxed during the Early - Mid Miocene, followed by

Late Miocene (Late Tortonian) uplift of the northern margin of the Caltanissetta Basin (Grasso and Pedley, 1988; Robertson and Grasso, 2005). To the south the Caltanissetta Basin remained deep until the onset of the MSC and its evaporite deposition and again deepened after it (Benson, 1973; Cita, 1973; Sprovieri et al., 1996a, b; Sgarrella et al., 1997), with estimated water depths of about 1200 m (Kouwenhoven et al., 2003; Kouwenhoven and Van der Zwaan, 2006; Krijgsman and Meijer, 2008). Although the palaeogeography of the basin changed continuously during the Messinian (Butler et al., 1995), after Rouchy and Caruso (2006) water depth of the central trough remained deep throughout this period until its uplift during the late Pliocene (Rouchy and Caruso, 2006).

The Caltanissetta Basin was tectonically separated in several thrust related synclines (sub-basins), resulting in different palaeobathymetric settings and a southward deepening of the basin (Butler et al., 1995, 1999; see Chapter 4, Fig. 1). The sedimentary sub-basins varied from proximal deltaic settings (e.g. Serra Pirciata, Torrente, Vaccarizzo, Marianopoli: Terravecchia Formation), with high sediment delivery and accumulation rates, to more distal pelagic to hemipelagic settings (eg. Falconara, Gibliscemi, Eraclea Minoa: Licata Formation).

In order to study the complete sedimentary record, preceding and following the MSC, we selected three sections, representative for the more distal and deeper parts of the basin: the Gibliscemi, Falconara and Eraclea Minoa section (see Chapter 4, Fig. 1). The Falconara–Gibliscemi composite section provides the complete history of the MSC, including the pre-evaporitic phase of the upper Tortonian to the onset of the MSC at 5.96 Ma (Blanc-Valleron et al., 2002). The Eraclea Minoa section comprises the complete succession of the MSC and the following early Pliocene Trubi marls with a continuous transition across the Miocene–Pliocene boundary. The succession of the MSC itself (evaporitic and Lago Mare phase) is not relevant for our study, as it contains no calcareous dinoflagellates. The same holds for the succession of the Pissouri Basin.

Sedimentary succession

The Neogene stratigraphic succession of Sicily includes the pre-evaporitic, the MSC and post-MSC sequence. The sedimentary succession, preceding and following the MSC, generally consist of cyclic alternations of indurated calcareous beds, softer marls, sapropels and laminated diatomites, developed in response to astronomical climate forcing (e.g. Hilgen et al., 1995, 1996; Sproveri et al., 1996a; Vazquez et al., 2000). The palaeoclimatic origin of the organic rich sapropels (> 2% Corg) is not fully understood yet, but they are commonly interpreted as the result of dominantly precession controlled dry–wet oscillations in the

Mediterranean realm (Hilgen and Krijgsman, 1999). Herewith, the sapropels correspond to precession minima and northern hemisphere (NH) summer insolation and humidity maxima (e.g. Rossignol-Strick, 1983; Hilgen, 1991; Hilgen and Krijgsman, 1999; Gallego et al., 2010). Usually sapropel formation is interpreted as the result of increased productivity due to enhanced continental runoff (Calvert et al., 1992; Diester-Haass et al., 1998; Lourens et al., 1992; Martinéz-Ruiz et al., 2000, 2003; Meyers and Arnaboldi, 2005) together with improved preservation of organic matter due to deep water oxygen depletion (Gallego et al., 2010). The origin of the diatomites is suggested to be astronomically controlled as well (Hilgen and Krijgsman, 1999). Usually its occurrence in the Mediterranean realm is interpreted as a first discrete expression of basin restriction and sea level fluctuation toward the MSC (e.g. McKenzie et al., 1979; Meulenkamp et al., 1979; Gersonde and Schrader, 1984; Thunell et al., 1984; Van der Zwaan and Gudjonsson, 1986; Grasso et al., 1991; Suc et al., 1995; Hüsing et al., 2009). The exact genesis of the diatomites is still unclear. Several authors propose a relation with upwelling, sea-level rise and increased runoff (e.g. McKenzie, 1979; Martin et al., 1991; Suc et al. 1995; Van Assen et al., 2006).

Lithological units

The pre-evaporitic succession is represented by the Gibliscemi-Falconara composite section, which comprises two main lithological units, the Licata and the Tripoli Formation (see Chapter 4, Fig. 2). It starts with the Licata Formation (Gibliscemi section; upper Tortonian– lower Messinian), which consists of sapropel - bearing marls of bipartite cycle, representative for more distal and deeper basin settings (Hilgen and Krijgsman, 1999). The Licata Formation represents the distal equivalent of the Terravecchia Formation, which consists of siliciclastic sediments of more proximal and shallower, normal marine basin positions.

The Tripoli Formation starts with the occurrence of the first diatomite bearing bed (Falconara section) (Hilgen and Krijgsman, 1999). The complete Tripoli Formation consists of 49, usually tripartite, precession cycles, consisting of reddish to brown sapropels, white diatomites and homogenous greenish marls. Its upward evolution reflects increasing restriction of the Mediterranean basin (Pedley and Grasso, 1993). On top of the Tripoli Formation starts the evaporite bearing Gessoso Solfifera Formation (gypsum – sulphur formation), whose basal part (Calcare di Base) marks the onset of the MSC.

The Gessoso Solfifera Formation consists of two main units, the Lower and Upper Evaporites (Decima and Wezel, 1973; Butler et al., 1995), which are separated by an erosional surface (MES) (Garcia-Veigas et al., 1995; Rouchy and Caruso, 2006). The Lower Evaporites comprise the massive evaporitic carbonates of the Calcare di Base, with pseudomorphs after halite and gypsum (Butler et al., 1999), which are followed by a variety of evaporate facies (e.g. potassium salts). The Upper Evaporites include thick detrital layers, which are associated with increased runoff (Decima and Wezel, 1973; Butler et al., 1995). The Messinian evaporites in the Caltanissetta Basin developed during a regressive-transgressive cycle (Butler et al., 1999). The Upper evaporites are overlaid by the siliciclastic Arenazzolo Formation, which contains a typical freshwater fauna, occurring in the latest phase of the MSC and indicating an important dilution episode, equivalent to the Lago Mare facies (Bonaduce and Sgarrella, 1999).

The MSC ends with a sharp transition to the deep marine deposits of the Trubi Formation at the base of the Pliocene (Zanclean) (e.g. Kastens and Mascle, 1990; Van Couvering et al., 2000; Iaccarino et al., 1999).

Eastern Mediterranean basin

The eastern Mediterranean basin and its surrounding land areas are remnants of the mainly Mesozoic Neotethys ocean, formed along the northern passive margin of the African plate during the early Mesozoic (Garfunkel, 1998). With the beginning of the Permian–Triassic period, rifting and continental breakup, a series of geological processes took place within the easternmost Mediterranean area, followed by passive margin subsidence and initial continental collision of the African and Eurasian plates along the active margin south of Cyprus (Robertson, 1998). Today, the boundary of the African and Eurasian plates runs almost east-west across the eastern Mediterranean Sea and is located between southern Cyprus and the Eratosthenes Seamount (Kempler and Ben-Avraham, 1987; Anastasakis and Kelling, 1991; Emeis et al., 1996; Robertson, 1998). Accordingly, the Mesozoic ophiolites and related continental margins in southern Turkey and Cyprus represent tectonically emplaced remnants of a southerly Neotethyan oceanic basin (Robertson, 1998). The Eratosthenes Seamount represents a rifted continental fragment, composed of a carbonate platform, associated with rift-related intrusions or extrusions of basic igneous rocks (Robertson, 1998 and citations herein). Currently it is in a dynamic collision phase with the active southern margin of Cyprus. The North African and Levantine coastal and offshore areas represent a passive continental margin of Triassic age, which passes oceanwards into oceanic crust (Bien and Gvirtzman, 1977; Garfunkel, 1998; Robertson, 1998). From the Late Cretaceous to Holocene the Eastern Mediterranean has been in a diachronous collisional phase (Robertson, 1998). Nowadays the eastern areas of the Levantine basin are in a post-collisional phase (Pearce et al., 1990), whereas the western areas are still in an early collisional phase (Robertson and Grasso, 1995).

Cyprus/Pissouri Basin

Cyprus is located in the easternmost Mediterranean Sea. It is divided into four main lithological units, which are from the north to the south, the Kyrenia Range, the Mesaoria Basin the Troodos ophiolite complex and the Mamonia complex (see Chapter 3, Fig. 1). The Kyrenia Range presents a fold – thrust belt, evolved during the Palaeocene, the Mesaoria Basin corresponds to an intra – montane settin, which evolved during the Neogene between the Troodos ophiolite and the Kyrenia range (Calon et al., 2005) and the Mamonia complex represents a fragment of an Mesozoic continental margin (Robertson, 1998).

The sedimentary succession on Cyprus is affected by a complex uplift history, which is attributed to its unstable position in a fore-arc setting of the convergence zone between Africa and Eurasia (e.g. Robertson, 1998; Orszag-Sperber et al., 1989). The tectonics first involved genesis of the Troodos massif (Robertson, 1977), a fragment of an old oceanic crust of late Cretaceous age which represents the Basement of Cyprus (Moores and Vine, 1971; Allerton and Gomez, 1989). This primary phase is followed by deformation processes in Maastrichtian time, the most active tectonic period in this region (Robertson, 1977; Orszag-Sperber et al., 1989). Then, after a brief period of latest Cretaceous deep-water pelagic sedimentation much of the area was covered in early Tertiary time by a wedge of pelagic carbonates derived from the northeast (Robertson and Hudson, 1974). Gradual uplift to the middle Miocene is documented by shallowing-upward carbonate sequences, culminating in the generation of local reef complexes and lagoons in adjacent areas of the Troodos complex (Robertson, 1977). Vastly accelerated middle Miocene uplift is recorded by slumping, widespread erosion, and folding of the sedimentary sequences south of Troodos (Robertson, 1977). The uplift was associated with rapid deposition of both clastic and carbonate sediments in subsiding basins to the south.

Four sedimentary sub-basins developed in this area during Miocene time, the Polemi, Pissouri, Limassol and Psematismenos basins (Orszag-Sperber et al., 1989; Robertson et al., 1995; Stow et al., 1995; Krijgsman et al., 2002). The main uplift phase of the Troodos is attributed to the early Messinian (Orszag-Sperber et al., 1989). In contrast, much of the Troodos massif remained low-lying and relatively stable during the Pliocene (Orszag-Sperber et al., 1989). The Messinian uplift phase was superimposed by sea level fall, caused by the progressive restriction and partial drying out of the Mediterranean Basin related to the MSC (Hsü et al., 1978) culminating in local reef development, gypsum deposition and wide-ranging erosion of the Troodos massif during the climax of the MSC (Robertson, 1998).

The evaporite deposition in the Pissouri Basin was contemporaneous with other evaporites of the Mediterranean realm and thus a supra-regional event (Orszag-Sperber et al., 1989). The gypsum deposits of the basin are interpreted as being entirely subaqueous with thicknesses of about 40 m in the central parts (Stow et al., 1995). Features for a complete drying out are only identified in some bed deposits directly overlying the evaporites (Stow et al., 1995). The evaporative phase was followed by a brackish episode "Lago Mare" at the end of the Messinian stage before the restoration of open marine conditions took place with the beginning of the Pliocene (Pierre et al., 1998; Rouchy et al., 2001). The post evaporitic Lago Mare phase was probably accompanied by a rapid fault - controlled subsidence of the Pissouri Basin (Stow et al., 1995). Much of the eroded terrigenous material from the Troodos ophiolite was discharged into the Pissouri Basin (Stow et al., 1995). Marine deep water conditions were reestablished with the beginning of the Pliocene (Stow et al., 1995; Rouchy et al., 2001; Krijgsman et al., 2002). Renewed shallowing upward appeared during the late Pliocene to early Pleistocene, associated with major uplift of those parts which today represent the island Cyprus (McCallum and Robertson, 1990; Stow et al., 1995).

For our study of the eastern Mediterranean we used sample material from the pre-evaporitic (late Tortonian/Messinian) and post-Messinian sequence (early Pliocene), which provides the complete sedimentary record preceding and following the MSC (see Chapter 2, Fig. 2; Chapter 3, Fig. 2).

Sedimentary succession

The Pissouri Basin on Cyprus contains one of the most suitable sedimentary sequences for studying the onset of the MSC in the Eastern Mediterranean (Krijgsman et al., 2002). It provides one of the most continuous sedimentary successions related to the MSC, comprising the pre-evaporitic, evaporitic and post-evaporitic deposits which cover the period from the late Tortonian up to the early Pliocene (Krijgsman et al., 2002; Orszag-Sperber et al., 2009). The section shows the typical sedimentary cyclicity of late Neogene Mediterranean sequences, comprising 48 distinct cycles (1–48) and 10 less pronounced cycles (I–VII, 3 x ?), associated with orbitally driven climate fluctuations (see Chapter 2, Fig. 2). Micropalaeontological, magneto- and cyclosratigraphic studies in the Pissouri Basin provide an excellent chronological framework of the timing and palaeoenvironmental changes that affected this basin (Krijgsman et al., 2002; Merle et al., 2002; Kouwenhoven et al., 2006). However, several sedimentary features indicate that tectonic processes may have affected the sedimentation in Cyprus basins, even if this local tectonics were not the main reason for the basin restriction (Orszag-Sperber et al., 2009).

Lithological units

The pre-evaporitic series starts (upper Tortonian/lower Messinian: cycle VII-42) with predominantly homogenous bluish marls, which are interbedded by thin sandy and carbonate layers in its upper part (Krijgsman et al., 2002) (see Chapter 3, Fig. 2). A distinct sedimentary cyclicity is missing in the lower part or is only expressed by color alternations between bluish marls and darker beds (Krijgsman et al., 2002). Upwards the cyclicity becomes more distinct by alternations between indurated carbonates and less indurated marls, which are combined to individual cycles of bipartite pattern (Krijgsman et al., 2002; Orszag-Sperber et al., 2009). This unit can be regarded as the more or less stratigraphic equivalent of the Licata Formation on Sicily. Contrary to this, the deposits from the Pissouri Basin represent an intermediate water depth of ~300 to 500 m around the Tortonian-Messinian boundary (Kouwenhoven et al., 2006).

In the following succession (upper Messinian: cycle 41–1), alternations of whitish indurated homogenous carbonates, marls and laminated marls occur, together with intercalations of dark laminated organic rich clayey layers, which are referred to as sapropels (Krijgsman et al., 2002). In the upper part (cycle 22–1), inclusions of gypsum and its pseudomorphs occur with an upward trend to increase. The individual cycles are bi- and tripartite. The thickness of the sapropelic layers is strongly reduced by bedding parallel shear planes (Krijgsman et al., 2002). Pure diatomites do not occur, merely softer laminated marls with a diatomitic appearance (Krijgsman et al., 2002). A thick (~3.5 m) slumped element, above a series of organic rich layers, interrupts the cyclic bedding. It consists of a chaotic accumulation of marl, diatomite and carbonate fragments (Krijgsman et al., 2002). The water depth is estimated to have been between about 500 and 200 m in the lower part and less than 100 m in the upper part (Krijgsman et al., 2002; Merle et al., 2002). After Kouwenhoven et al. (2006) the water depth remained rather stable around 300 – 500 m until the onset of the MSC. However, this interval correlates to the Tripoli Formation on Sicily and distinctly reflects progressive palaeoenvironmental changes, caused either by shallowing or by other unfavourable environmental conditions (Orszag-Sperber et al., 2009).

This sequence is followed by a stromatolite-bearing series, the so called "barre jaune", which marks the beginning of the MSC in Cyprus basins at 5.96 Ma (Orszag-Sperber et al., 2009). It can be correlated with the Calcare di Base unit on Sicily and like this it represents a transitional interval from the pre-evaporitic cyclic carbonates to the evaporitic succession with the gypsum deposits (Orszag-Sperber et al., 2009). The evaporitic phase comprises two gypsum units, lower and upper Gypsum (equivalent to the Lower and Upper Evaporites on Sicily) separated by a breccia (Orszag-Sperber et al., 1980, 2009; Rouchy et al., 2001; Krijgsman et al., 2002), which can be considered as an equivalent to the Messinian erosional surface (MES). The onset of brackish water conditions "Lago Mare" facies - interrupted by periods of subaerial exposure and increased marine influences - terminates the evaporitic phase of the MSC (Rouchy et al., 2001).

Marine marls of variable colours (Trubi Formation; early Pliocene) overlie the Lago Mare deposits with a sharp contact and indicate the re-establishment of fully marine conditions at the beginning of the Pliocene (Zanclean) and thus the end of the MSC in the Pissouri Basin (Rouchy et al., 2001). A water depth of at least 300 m was estimated by Rouchy et al., (2001).

References

Accaino, F., Catalano, R., Di Marzo, L., Giustiniani, M., Tinivella, U., Nicolich, R., Sulli, A., et al., 2010 (in press). A crustal seismic profile across Sicily. *Tectonophysics*: doi:10.1016/j.tecto.2010.07.011.

Adams, C.G., Gentry, A.W., Whybrow, P.J., 1983. Dating the terminal Tethyan event. In: Meulenlkamp, J.E. (Ed.). Reconstruction of Marine Paleoenvironments. *Micropaleontological Bulletin*, Utrecht, **30**: 273–298.

Addadi, L., Weiner, S., 1989. Stereochemical and structural re- lations between macromolecules and crystals in hiomineralization. In: Mann, S., Webb, J., Williams, R.J.P. (Eds.). *Biomineralization*, VCH, Weinheim: 133-56.

Addadi, L., Berman, A, Moradian-Oldak, J., Weiner, S., 1990. Tuning of crystal nucleation and growth by proteins: molecular interactions at solid-liquid interfaces in hiomineralization. *Croatica Chemica Acta*, **63**: 539-44.

Aguirre, J., Sánchez-Almazo, I.M., 2004. The Messinian post- evaporitic deposits of the Gafares area (Almería-Níjar basin, SE Spain). A new view of the "Lago-Mare" facies. *Sedimentary Geolohy*, **168**: 71–95.

Allerton, S., Gomez, B., 1989. Structural Geomorphology of Southeast Troodos, in the Vicinity of Pano Lefkara, Cyprus. Geografiska Annaler. *Physical Geography*, **71**(A): 221.

Anastasakis, G., Kelling, G., 1991. Tectonic connection of the Hellenic and Cyprus arcs and related geotectonic elements. *Marine Geology*, **97**: 261– 277.

Anderson, D.M.; Taylor, C.D., Armbrust, E.V., 1987. The effects of darkness and anaerobiosis on dinoflagellate cyst germination. *Limnology and Oceanography*, **32**(2): 340 – 351.

Barber, P.M., 1981. Messinian subaerial erosion of the Proto-Nile delta. *Marine Geology*, **44**: 253–272.

Barbieri R, Ori, G.G., 2000. Neogene palaeoenvironmental evolu- tion in the Atlantic side of the Rifian Corridor (Morocco). *Palaeogeography Palaeoclimatology Palaeoecology*, **163**: 1–31.

Barr, F.T., Walker, B.R., 1973. Late Tertiary channel system in Northern Libya and its implications on Mediterranean sea level changes. In: Ryan, W.B.F., Hsü, K.J. et al. (Eds.). *Initial Reports of the Deep Sea Drilling Project*, **13**: 1244-1251.

Baumann, K.-H., Böckel, B., Donner, B., Gerhardt, S., Henrich, R., Vink, A., Volbers, A, Willems, H., Zonneveld, K.A.F., 2003. Contribution of calcareous plankton groups to the carbonate budget of South Atlantic surface sediments. In: Wefer, G., Mulitza, S., Ratmeyer, V. (Eds.). The South Atlantic in the Late Quaternary: reconstruction of material budget and current systems. *Springer Verlag*, Berlin, Heidelberg: 81-99.

Benson, R.H., 1973. An ostracodal view of the Messinian salinity crisis. In: Drooger, C.W. (Ed.). Messinian events in the Mediterranean. *Geodynamics Scientific Report*, Koninklijke Akademie van Wetenschappen, Amsterdam, **7**: 235-242.

Benson, R.H., Rakic-El Bied, K., Bonaduce, G., 1991. An important current reversal (influx) in the Rifian corridor (Morocco) at the Tortonian-Messinian boundary: the end of Tethys Ocean. *Paleoceanography*, **6**: 164–192.

Benson, R.H., Hayek, L. C., Hodell, A., 1995. Extending the climatic precession curve back into Late Miocene by signature template comparison. *Paleoceanography*, **10**(1): 5-20.

Beranger, K., Mortier, L., Crepon, M., 2005. Seasonal variability of water transport through the Straits of Gibraltar, Sicily and Corsica, derived from a high-resolution model of the Mediterranean circulation. *Progress In Oceanography*, **66**: 341-364.

Bertini, A, 2006. The Northern Apennines palynological record as a contribute for the reconstruction of the Messinian palaeoenvironments. *Sedimentary Geology*, **188-189**: 235-258.

Bethoux, J.P., Gentili, B., 1999. Functioning of the Mediterranean Sea: past and present changes related to freshwater input and climate changes. *Journal of Marine Systems*: 33-47.

Bertoni, C., Cartwrigth, A., 2006. Controls on the basinwide architecture of late Miocene (Messinian) evaporites on the Levant margin (Eastern Mediterranean). Sedentary Geology, **188-189**: 93-114.

Bien, A., Gvirtzman, G., 1977. A Mesozoic fossil edge of the Arabian plate along the Levant coastline and its bearing on the evolution of the Eastern Mediterranean. In: Biju-Duval, B., and Montadert, L. (Eds.). Structural History of the Mediterranean Basins. International Symposium Split (Yugoslavia), Oct. 1976. *Edition Technip*, Paris: 95–110.

Binder, B.J., Anderson, D.M., 1987. Physiological and environmental control of germination of Scrippsiella trochoidea (Dinophyceae) resting cysts. *Journal of Phycology*, **23**: 99–107.

Binder, BJ, Anderson, D.M, 1990. Biochemical composition and metabolic activity of *Scrippsiella trochoidea* (Dinophyceae) resting cysts. *Journal of Phycology*, **26**: 289–298.

Bison, K.-M., 2001. Verbreitung kalkiger Dinoflagellatenzysten im Schreibkreideprofil von Lägerdorf / Kronsmoor (oberes Ober- Campan bis unteres Unter-Maastricht). *Diplomarbeit*, Universität Bremen, Germany: 200pp.

Bison, K.-M., Versteegh, G.J.M., Hilgen, F.J., Willems, H., 2007. Calcareous dinoflagellate turnover in relation to the Messinian salinity crisis in the eastern Mediterranean Pissouri Basin, Cyprus. *Journal of Micropalaeontology*, **26**: 103-116.

Bison, K.-M., Versteegh, G.J.M., Orszag-Sperber, F., Rouchy, J.-M., Willems, H., 2009. Palaeoenvironmental changes of the early Pliocene (Zanclean) in the eastern Mediterranean Pissouri Basin (Cyprus) evidenced from calcareous dinoflagellate cyst assemblages. *Marine Micropaleontology*, **73**: 49-56.

Blanc, P.-L., 2000. Of sills and straits: a quantitative assessment of the Messinian Salinity Crisis. *Deep Sea Research*, **47**: 1429–1460.

Blanc, P.-L., 2006. Improved modelling of the Messinian Salinity Crisis and conceptual implications. *Palaeogeography, Palaeoclimatology, Palaeoecology*, **238**: 349-372.

Blanc-Valleron, M.-M., Rouchy, J.-M., Pierre, C., Badaut-Trauth, D., Schuler, M., 1998. Evidence of Messinian non-marine deposition at site 968 (Cyprus Lower Slope). *Proceedings of the Ocean Drilling Program, Scientific Results*, College Station, TX, **160**: 437-445.

Blanc-Valleron, M.-M., Pierre, C., Caulet, J.P., Caruso, A., Rouchy, J.-M., Cespuglio, G., Sprovieri, R., Pestrea, S., Di Stefano, E., 2002. Sedimentary, stable isotope and micropaleontological records of paleoceanographic change in the Messinian Tripoli Formation (Sicily, Italy). *Palaeogeography, Palaeoclimatology, Palaeoecology*, **185**: 255- 286.

Bolle, H.-J., 2003. Climate, Climate Variability, and Impacts in the Mediterranean Area: An Overview. In: Bolle, H.-J. et al. (Ed.). Mediterranean Climate, Variability and Trends. Regional Climate Studies. *Springer Verlag*, Berlin Heidelberg, Germany: 372pp.

Bolli, H.M., 1974. Jurassic and Cretaceous Calcisphaerulidae from DSDP Leg 27, Eastern Indian Ocean. In: Veevers, J.J., Heirtzler, J.R. et al. (Eds.). *Initial Reports of the Deep Sea Drilling Project*, **27:** 843-907.

Bonaduce, G., Sgarrella, F., 1999. Paleoecological interpretation of the latest Messinian sediments from southern Sicily (Italy). *Memorie della Società Geologica Italiana*, **54**: 83–91.

Bouchet, P., Taviani, M., 1992**.** The Mediterranean deep-sea fauna: pseudopopulations of Atlantic species? *Deep-sea Research*, **39**: 169-184.

Bouillon, J., Medel, M.D., Pagos, F., Gili, J.-M., Boero, F., Gravili, C., 2004. Fauna of the Mediterranean Hydrozoa. *Scientia Marina* (Barcelona), *Consejo Superior de Investigaciones Cientificas*, Barcelona, Spain, **68**(2): 5-449.

Burkholder, J.M., Glasgow, H.B., Deamer-Melia, N.J., 2001. Overview and present status of the toxic Pfiesteria complex. *Phycologia*, **40**: 186–214.

Butler, R.W.H., Grasso, M., Lickorish, H., 1995. Plio-Quaternary megasequence geometry and its tectonic controls within the Maghrebian thrust belt of south- central Sicily. *Science*, **7**: 171-178.

Butler, R.W.H., McClelland, E., Jones, R.E., 1999. Calibrating the duration and timing of the Messinian salinity crisis in the Mediterranean: linked tectonoclimatic signals in thrust-top basins in Sicily. *Journal of the Geological Society*, London, **156**: 827-835.

Bütschli, O., 1885. Protozoa. In: Bronn, H.G. (Ed.). Klassen und Ordnungen des Thier-Reichs, wissenschaftlich dargestellt in Wort und Bild. *1. Wintersche Verlagshandlung*, Leipzig: 865-1088.

Calon, T., Aksu, Hall, J., 2005. The Oligocene-Recent evolution of the Mesaoria Basin (Cyprus) and its western marine extension, Eastern Mediterranean. *Marine Geology*, **221**: 95-120.

Calvert, S.E., Nielsen B., Fontugne, M.R., 1992. Evidence from nitrogen isotope ratios for enhanced productivity during formation of eastern Mediterranean sapropels. *Nature*, **359**: 223-225.

Catalano, R., Di Stefano, P., Sulli, A., Vitale, F., 1996. Paleogeography and structure of the central Mediterranean: Sicily and its offshore area. *Tectonophysics*, **260**: 291-323.

Cavazza, W., Roure, F.M., Ziegler, P.A., 2004. The Mediterranean Area and the Surrounding Regions: Active Processes, Remnants of former Tethyan Oceans and Related Thrustbelts. In: Cavazza, W., Roure, F.M., Spakman, W., Stampfli, G.M., Ziegler, P.A. (Eds.). The TRANSMED Atlas. The Mediterranean Region from Crust to Mantle. *Springer Verlag*, Berlin–Heidelberg: 53–90.

Chumakov, I.S., 1973. Pliocene and Pleistocene deposits of the Nile Valley in Nubia and Upper Egypt. *Initial Reports of the Deep Sea Drilling Project*, **13**: 1242–1243.

CIESM (Roveri, M., Krijgsman, W., Suc, J.-P., Lugli, S., Lofi, J., Sierro, F.J., Manzi, V., Flecker, R., and others.), 2008. Executive Summary. In: Briand, F., (Ed.). The Messinian Salinity crisis from mega-deposits to microbiology - A consensus report. *CIESM Workshop Monographs*, **33**: 7-28.

Cita, M.B., Gartner, S., 1973. The Stratotype Zanclean: foraminiferal and nannofossil biostratigraphy. *Rivista Italiana di Paleontologia i Stratigrafia*, **79**: 503– 558.

Cita, M.B., Wright, R.C., Ryan, W.B.F., Longinelli, A., 1978. Messinian paleoenvironments. In: Hsü, K.J., Montadert, L., et al. (Eds.). *Initial Reports of the Deep Sea Drilling Project*, **42**(1): 1003-1035.

Clauzon, G., Rubino, J.-L., Savoye, B., 1995. Marine Pliocene Gilbert- type fan deltas along the French Mediterranean coast. A typical infill feature of preexisting subaerial Messinian canyons. In: IAS- 16th Regional Meeting of Sedimentology, *Field Trip Guide Book*, ASF Edition, Paris, **23**: 145–222.

Clauzon, G., Suc, J.-P., Gautier, F., Berger, A., Loutre, M.F., 1996. Alternate interpretation of the Messinian salinity crisis: controversy resolved? *Geology*, **24**: 363–366.

Clegg, M.R., Maberly, S.C., Jones, R.I., 2007. Behavioural response as a predictor of seasonal depth distribution and vertical niche separation in freshwater phytoplanktonic flagellates. *Priest*, **52**: 441-455.

Cubito, A., Ferrara, V., Pappalardo, G., 2005. Landslide hazard in the Nebrodi Mountains (Northeastern Sicily). *Geomorphology*, **66**: 359-372.

Cramp, A., Collins, M.B., West, R., 1988. Late Pleistocene–Holocene sedimentation in the NW Aegean Sea: a palaeoclimatic palaeoceanographic reconstruction. *Palaeogeography, Palaeoclimatology, Palaeoecology*, **68**: 61–77.

Dale, B., 1983. Dinoflagellate resting cysts: 'benthic plankton'. In: Fryxell, G.A., (Ed.). Survival Strategies of the Algae. *Cambridge University Press,* Cambridge: 69-137.

Dale, B., 1992a. Dinoflagellate contribution to the open ocean sediment flux. In: Honjo, S. (Ed.). Dinoflagellate contribution to the Deep Sea. *Ocean Biocoenosis Series*, Oceanographic Institution, Woods Hole, Massachusetts, **5**: 1-32.

Dale, B., 1992b. Thoracosphaerids: Pelagic Fluxes. In: Honjo, S. (Ed.). Dinoflagellate Contributions to the Deep Sea. *Ocean Biocoenosis Series*, Oceanographic Instituion, Woods Hole, Massachusetts, **5**: 33-44.

Decima, A., Wezel, F. C., 1973. Late Miocene evaporites of the Central Sicilian Basin, Sicily.- In: Ryan, W.B.F., Hsü, K.J. et al.: Initial Rep. *Deep-Sea Drilling Project*, Washington, D.C., U.S. Government Printing Office, **13**(2): 1234-1240,

Deflandre, G., 1947. *Calciodinellum* nov. gen., premier representant d' une famille nouvelle de Dinoflagellates fossiles a theque calcaire. *Comptes Rendus Académie des sciences*, **224**: 1781-1782.

Dias-Brito, D., 2000. Global stratigraphy, palaeobiogeography and palaeoecology of Albian–Maastrichtian pithonellid calcispheres: impact on Tethys configuration. *Cretaceous Research*, **21**: 315-349.

Diester-Haass, L., Robert, C., Chamley, H., 1998. Paleoproductivity and climate variations during sapropel deposition in the eastern Mediterranean Sea. *Proceedings of the Ocean Drilling Program, Scientific Results,* **160**: 227-248.

Duggen, S., Hoernle, K., Boogard, P.V.D., Rupke, L., Phipps Morgan, J. (2003). Deep roots of the Messinian salinity crisis. *Nature*, **422**: 602-606.

Elbrächter, M., Gottschling, M., Hildebrand-Habel, T., Keupp, H., Kohring, R., Lewis, J., Meier, S.K.J., Montresor, M., Streng, M., Versteegh, G.J.M., Willems, H., Zonneveld, K.A.F., 2008. Establishing an

Agenda for Calcareous Dinoflagellate Research (Thoracosphaeraceae, Dinophyceae) including a nomenclatural synopsis of generic names. *Taxon*, **57**: 1289-1303.

Emeis, K.-C., Sakamoto, T., Wehausen, R., Brumsack, H.-J., 2000. The sapropel record of the Eastern Mediterranean Sea – results of Ocean Drilling Program Leg 160. *Palaeogeography, Palaeoclimatology, Palaeoecology*, 158, 371–395.

Esper, O., Versteegh, G.J.M., Zonneveld, K.A.F., Willems, H., 2004. A palynological reconstruction of the Agulhas Retroflection (South Atlantic Ocean) during the Late Quaternary. *Global and Planetary Change*, **41**: 31-62.

Evitt, W.R., 1967. Dinoflagellate studies II. The archaeopyle. Stanford University Publications, *Geological Science*, **10**(3): 1-83.

Evitt, W.R., 1985. Sporopollenin dinoflagellate cysts: their morphology and interpretation. *AASP Monograph Series*: 333 S.

Favre, E., Francois, L., Fluteau, F., Cheddadi, R., Thevenod, L., Suc, J., 2007. Messinian vegetation maps of the Mediterranean region using models and interpolated pollen data. *Geobios*, **40**: 433-443.

Garcia-Veigas, J., Orti, F., Rosell, L., Ayora, C., Rouchy, J.-M., Lugli, S., 1995. The Messinian salt of the Mediterranean: geochemical study of the salt from the Central Sicily Basin and comparison with the Lorca Basin (Spain). *Bulletin Societé Géologique*, **166**: 699-710.

Garfunkel, Z., 1998. Constrains on the origin and history of the Eastern Mediterranean basin. *Tectonophysics*, **298**: 5-35.

Butler, R.W.H, Grasso, M., 1993. Tectonic controls on base-level variation and depositional sequences within thrust-top and foredeep basins: examples from the Neogene thrust belt of central Sicily. *Basin Research*, **5**: 137–151.

Fütterer, D., 1977. Distribution of calcareous dinoflagellates in Cenozoic sediments of Site 366, Eastern North Atlantic. *Initial Reports of the Deep Sea Drilling Project*, **41**: 533-541.

Gaines, G., Elbrächter, M., 1987. Heterotrophic nutrition In: Taylor, F.J.R. (Ed.). The biology of dinoflagellates. *Blackwell*, Oxford: 224-268.

Gallego-Torres, D., Martínez-Ruiz, F., Meyers, P. a, Paytan, a, Jimenez-Espejo, F.J., Ortega-Huertas, M., 2010. Productivity patterns and N-fixation associated with Pliocene-Holocene sapropels: paleoceanographic and paleoecological significance. *Biogeosciences Discussions*, **7**: 4463-4503.

Garfunkel, Z., 1998. Constrains on the origin and history of the Eastern Mediterranean basin. *Tectonophysics*, **298**: 5-35.

Gersonde, R. and Schrader, H., 1984. Marine planktic diatom correlation of lower Messinian deposits in the Western Mediterranean. Marine Micropalaeontology, **9**: 93-110.

Gilbert, M.W., Clark, D.L., 1983. Central Arctic Ocean paleooceanographic interpretation based on Late Cenozoic calcareous dinoflagellates. *Marine Micropaleontology*, **7**: 385-40l.

Gottschling, M., Plötner, J., 2004. Secondary structure models of the nuclear internal transcribed spacer regions and 5.8S rRNA in Calciodinelloideae (Peridiniaceae) and other dinoflagellates. *Nucleic Acids Research*, **32**(1): 307-315.

Gottschling, M., Keupp, H. Plötner, J., Knop, R., Willems, H., Kirsch, M., 2005. Phylogeny of calcareous dinoflagellates as inferred from IST and ribosomal sequence data. *Molecular Phylogenetics and Evolution*, **36**: 444-455.

Govers, R., Meijer, P., Krijgsman, W., 2009. Tectonophysics Regional isostatic response to Messinian Salinity Crisis events. *Tectonophysics*, **463**: 109-129.

Grasso, M., Butler, R.W.H., La Manna, F., 1991. Thin skinned deformation and structural evolution in the NE segment of the Gela Nappe, SE Sicily. In: Boccaletti, M., Decima, G., Papani, G. (Eds.). Neogene Thrust Tectonics. *Studi Geologia*, Carneri, Parma: 9-17.

Griffin, D.L., 1999. The late Miocene climate of northeastern Africa: unravelling the signals in the sedimentary succession. *Journal of the Geological Society, London*, **156**: 817-826.

Griffin, D.L., 2002. Aridity and humidity: two aspects of the late Miocene climate of North Africa and the Mediterranean. *Palaeogeography, Palaeoclimatology Palaeoecology*, **182**: 65–91.

Hallegraeff, G.M., C. Bolch., 1992. Transport of diatom and dinoflagellate resting spores in ships' ballast water: implications for plankton biogeography and aquaculture. *Journal of Plankton Research*, **14**: 1067–1084.

Hardie, L.A., Lowenstein, T.K. Did the Mediterranean Sea dry out during the Miocene? A reassessment of the evaporite evidence from DSDP legs 13 and 42A cores. *Journal of Sedimentary Research*, **74**(4): 453pp.

Harland, R., 1983. Distribution maps of recent dinoflagellate cysts in bottom sediments from the North Atlantic Ocean and adjacent seas. *Palaeontology*, **26**: 321-387.

Head, M.J., 1996. Modern dinoflagellate cysts and their biological affinities. In: Jansonius, J., McGregor, D.C. (Eds*). Palynology: Principles and Applications,* AASP Foundation, Dallas, TX, **3**: 1197-1248.

Hildebrand-Habel, T., Willems, H., 1997. Calcareous dinoflagellate cysts from the Middle Coniacian to Upper Santonian chalk facies of Lägerdorf (N Germany). *Courier Forschungsinstitut Senckenberg*, **201**: 177-199.

Hildebrand-Habel, T., Willems, H., 1999. New calcareous dinoflagellates from the Palaeogene of the South Atlantic Ocean (DSDP Site 357, Rio Grande Rise). 1. *Micropalaeontology*, **18**: 89-95.

Hildebrand-Habel, T., Willems, H., Versteegh, G.J.M., 1999. Variations in calcareous dinoflagellate associations from the Maastrichtian to Middle Eocene of the western South Atlantic Ocean (Sao Paulo Plateau, DSDP Leg 39, Site 356). *Review of Palaeobotany and Palynology*, **106**(1-2): 57-87.

Hildebrand-Habel, T., Willems, H., 2000. Distribution of calcareous dinoflagellates from the Maastrichtian to early Miocene of DSDP Site 357 (Rio Grande Rise, western South Atlantic Ocean). *International Journal of Earth Sciences*, **88**: 694-707.

Hildebrand-Habel, T., Streng, M., 2003. Calcareous dinoflagellate associations and Maastrichtian–Tertiary climatic change in a high latitude core (ODP Hole 689B, Maud Rise, Weddell Sea). *Palaeogeography, Palaeoclimatology, Palaeoecology,* **197**: 293–321.

Hilgen, F.J., 1991. Astronomical calibration of Gauss to Matuyama sapropels in the Mediterranean and implication for the geomagnetic polarity time scale. *Earth and Planetary Science Letter*, **104**: 226-244.

Hilgen, F.J., Krijgsman, W., 1999. Cyclostratigraphy and astrochronology of the Tripoli diatomite formation (pre-evaporite Messinian Salinity Crisis, Italy). *Terra Nova*, **11**: 16–22.

Hilgen, F.J., Krijgsman, W., Langereis, C.G., Lourens, L.J., Santarelli, A., Zachariasse, W.J., 1995. Extending the astronomical (polarity) time scale into the Miocene. Earth Planetary Science Letters, **136**: 495–510.

Hilgen, F.J., Krijgsman, W., Raffi, I., Turco, E., Zachariasse, W.J., 2000. Integrated stratigraphy and astronomical calibration of the Serravallian/Tortonian boundary section at Monte Gibliscemi (Sicily, Italy). *Marine Micropaleontology*, **38**: 181–211.

Hilgen, F., Kuiper, K., Krijgsman, W., Snel, E., Van der Laan, E., 2007. Astronomical tuning as the basis for high resolution chronostratigraphy: the intricate history of the Messinian Salinity Crisis. *Stratigraphy*, **4**: 231–238.

Hodell, D.A., Curtis, J.H., Sierro, J., Raymo, M.E., 2001. Correlation of late Miocene to early Pliocene sequences. *Atlantic*, **16**: 164-178.

Höll, C., Zonneveld, K.A.F., Willems, H., 1998. On the ecology of calcareous dinoflagellates: The Quaternary Eastern Equatorial Atlantic. *Marine Micropaleontology*, **33**(1-2): 1-25.

Höll, C., Kemle von Mücke, S., 2000. Late Quaternary upwelling variations in the eastern equatorial Atlantic Ocean as inferred from dinoflagellate cysts, planktonic foraminifera, and organic carbon content. *Quaternary Research*, **54**(1): 58-67.

Hsü, K.J., Cita, M.B., Ryan, W.B.F., 1973. The origin of the Mediterranean evaporites. *Initial Reports of the Deep Sea Drilling Project*, **13**: 1203–1231.

Hsü, K.J., Montadert, L., Bernoulli, D., Cita, M.B., Erickson, A., Garrison, R.E., Kidd, R.B., Mèlierés, F., Müller, C., Wright, R., 1977. History of the Mediterranean salinity crisis. *Nature*, **267**: 399-403.

Hsü, K.J., Stoffers, P., Ross, D.A., 1978. Messinian evaporites from the Mediterranean and Red Seas. *Marine Geology*, **26**(1-2): 71-72.

Hsü, K.J., 1978. The Messinian salinity crisis. *Naturwissenschaften (Historical Archive)*, **5**(3): 151.

Hsü, K.J., Bernoulli, D., 1978. Genesis of the Tethys and the Mediterranean. *Initial Reports of the Deep Sea Drilling Project*, **42**: 943-9.

Hüsing, S.K., Kuiper, K.F., Link, W., Hilgen, F.J., Krijgsman, W., 2009. The upper Tortonian–lower Messinian at Monte dei Corvi (Northern Apennines, Italy): Completing a Mediterranean reference section for the Tortonian Stage. *Earth and Planetary Science Letters*, **282**: 140–157.

Iaccarino, S.M., Castradori, D., Cita, M.B., Di Stefano, E., Gaboardi, S., McKenzie, J.A., Spezzaferri, S., Sprovieri, R., 1999. The Miocene-Pliocene boundary and the significance of the earliest Pliocene flooding in the Mediterranean. *Memorie della Societa Geologica Italiana*, **54**: 109-131.

Inouye, I., Pienaar, R.N., 1983. Observations on the life cycle and microanatomy of Thoracosphaera heimii (Dinophyceae) with special reference to its systematic position. *South African Journal of Botany*, **2**: 63–75.

Janofske, D., 1992. Kalkiges Nannoplankton, insbesondere kalkige Dinoflagellaten-Zysten der alpinen Ober-Trias: Taxonomie, Biostratigraphie und Bedeutung für die Phylogenie der Peridiniales. *Berliner geowississenschaftliche Abhandlungen*, **4**(E): 53

Janofske, D., Keupp, H., 1992. Mesozoic and Cenozoic ''calcispheres'' update in systematics. *International Nannoplankton Association Newsletter*, **14**: 14–16.

Janofske, D., 1996. Ultrastructure types in recent "Calcispheres". *Bulletin de l'Institut océanographique*, **14**(4): 295-303.

Janofske, D., 2000. Scrippsiella trochoidea and Scrippsiella regalis, nov. comb. (Peridiniales, Dinophyceae): a comparison. *Journal of Phycology*, **36**: 178–189.

Janofske, D., Karwath, B., 2000. Oceanic calcareous dinoflagellates of the equatorial Atlantic Ocean: Cyst-theca relationship, taxonomy and aspects on ecology. *In*: Karwath, 2000, Ecological Studies on Living and Fossil Calcareous Dinoflagellates of the Equatorial and Tropical Atlantic Ocean. *Berichte Fachbereich Geowissenschaften*, Universität Bremen, **152**: 94-136.

Jenner, R.A. , Littlewood, D.T.J., 2008. Problematica old and new. *Philosophical transactions of the Royal Society of London. Biological sciences*, **363**(B): 1503-1512.

Jiménez-Moreno, G., Fauquette, S., Suc, J.-P., 2009. Miocene to Pliocene vegetation reconstruction and climate estimates in the Iberian Peninsula from pollen data. *Review of Palaeobotany and Palynology*: 1-13. doi:10.1016/j.revpalbo.2009.08.001.

Jolivet, L., Augier, R., Robin, C., Suc, J., Rouchy, J., 2006. Lithospheric-scale geodynamic context of the Messinian salinity crisis. *Sedimentary Geology*, **188-189**: 9-33.

Karwath, B., Janofske, D., Tietjen, F., Willems, H., 2000a. Temperature effects on growth and cell size in the marine calcareous dinoflagellate *Thoracosphaera heimii*. *Marine Micropaleontology*, **39**(1-4): 43-51.

Karwath, B., Janofske, D., Willems, H., 2000. Spatial distribution of the calcareous dinoflagellate *Thoracosphaera heimii* in the upper water column of the tropical and equatorial Atlantic. *International Journal of Earth Sciences*, **88**: 668-679.

Keupp, H., 1987: Die kalkigen Dinoflagellatenzysten des Mittelalb bis Untercenoman von Escalles/Boulonnais (N-Frankreich). *Facies*, Erlangen **16**: 37- 88.

Keupp, H., Versteegh, G., 1989. Ein neues systematisches Konzept für kalkige Dinoflagellaten-Zysten der Subfamilie Orthopithonelloideae Keupp 1987. *Berliner geowissenschaftliche Abhandlungen*, Berlin, **106**(A): 207-219.

Keupp, H., 1981. Die kalkigen Dinoflagellaten-Zysten der borealen Unter-Kreide (Unter-Hauterivium bis Unter-Albium). *Facies*, **5**(S): 190.

Keupp, H., 1984. Revision der kalkigen Dinoflagellaten-Zysten G. Deflandres, 1948. *Paläontologische Zeitschriften,* **58**(1): 9-31.

Keupp, H., 1991. Fossil calcareous dinoflagellate cysts. In: Riding, R. (Ed.). Calcareous Algae and Stromatolites. *Springer Verlag,* Berlin, Heidelberg, New York: 267-286.

Keupp, H., 1993. Kalkige Dinoflagellaten-Zysten in Hell-Dunkel-Rhythmen des Ober-Hauterive/Unter-Baneme NW-Deutschlands. *Zitteliana*, **20**: 25-39.

Keupp, H., Kohring, R., 1993. Kalkige Dinoflagellatenzysten aus dem Obermiozän von El Medhi (Algerien). *Berliner Geowisswissenschaftliche Abhandlungen*, **9**: 25-43.

Keupp, H., 1995. Vertical distribution of calcareous dinoflagellate cysts of the Middle Aptian coresection Hoheneggelsen KB3 borehole, Lower Saxony, Germany. *Neues Jahrbuch, Geologisch Paläontologische Abhandlungen*, **196**(2): 221-233.

Keupp, H., Bellas, S.M., Frydas, D., Kohring, R. 1994. Aghia Irini, ein Neogenprofil auf der Halbinsel Gramvoússa/NW-Kreta. *Berliner Geowissenschaftliche Abhandlungen*, **13** (E): 469-481.

Keupp, H., Kohring, R. 1999. Kalkiger Dinoflagellatenzysten aus dem Obermiozän (NN 11) W von Rethymnon (Kreta). *Berliner Geowissenschaftliche Abhandlungen*, **30**(E): 33-53.

Keupp, H., Kowalski, F.-U., 1992. Die kalkigen Dinoflagellaten-Zysten aus dem Alb von Folkestone/SE-England. *Berliner geowissenschaftliche Abhandlungen,* **3**(E): 211-251.

Keupp, H., Mutterlose, J., 1984. Organismenverteilung in den D-Beds von Speeton (Unterkreide, England) unter besonderer Berücksichtigung der kalkigen Dinoflagellaten-Zysten. *Facies*, **10**: 153-178.

Keupp, H., Versteegh, G., 1989. Ein neues Konzept für kalkige Dinoflagellaten-zysten der Subfamilie Orthopithonelloideae Keupp 1987. *Berliner Geowissenschaftliche Abhandlungen*, **106**(A): 207-219.

Kienel, U., 1994. Die Entwicklung der kalkigen Nannofossilien und der kalkigen Dinoflagellaten- Zysten an der Kreide/Tertiär-Grenze in Westbrandenburg im Vergleich mit Profilen in Nordjütland und Seeland (DK). *Berliner geowississsenschaftliche Abhandlungen,* **12**(E): 87pp.

Kohring, R., 1993a. Kalkdinoflagellaten aus dem Mittel- und Obereozän von Jütland (Dänemark) und dem Pariser Becken (Frankreich) im Vergleich mit anderen Tertiär-Vorkommen. *Berliner Geowissenschaftliche Abhandlungen*, **6**(E): 1-164.

Kohring, R., 1993b. Kalkdinoflagellaten-Zysten aus dem unteren Pliozän von E-Sizilien. *Berliner Geowissenschaftliche Abhandlungen*, **9**(E): 15-23.

Kohring, R., 1997. Calcareous dinoflagellate cysts from the Blue Clay formation (Serravalian, Late Miocene) of the Maltese Islands. *Neues Jahrbuch, Geologisch-Paläontologische Mitteilungen*, **3**: 151-164.

Kohring, R., Gottschling, M., Keupp, H. 2005. Examples for character traits and palaeoecological significance of calcareous dinoflagellates. *Paläontologische Zeitschrift*, **79**(1): 79-91.

Kouwenhoven, T.J., Hilgen, F.J., van der Zwaan, G.J., 2003. Late Tortonian–early Messinian stepwise disruption of the Mediterranean–Atlantic connections: constraints from benthic foraminiferal and geochemical data. *Palaeogeography, Palaeoclimatology, Palaeoecology,* **198**: 303–319.

Kouwenhoven, T.J., Van der Zwaan, G.J., 2006. A reconstruction of late Miocene Mediterranean circulation patterns using benthic foraminifera. *Palaeogeography, Palaeoclimatology, Palaeoecology,* **238**: 373–385.

Kouwenhoven, T. J., Morigi, C., Negri, A., Giunta, S., Krijgsman, W., Rouchy, J.-M., 2006. Paleoenvironmental evolution of the eastern Mediterranean during the Messinian: Constraints from integrated microfossil data of the Pissouri Basin (Cyprus). *Marine Micropaleontology*, **60** (1): 17-44.

Krijgsman, W., Hilgen, F.J., Langereis, C.G., Santarelli, A., Zachariasse, W.J., 1995. Late Miocene magnetostratigraphy, biostratigraphy and cyclostratigraphy in the Mediterranean. *Earth and Planetary Science Letters*, **136**: 475-494.

Krijgsman, W., Langereis, C.G., Zachariasse, W.J., Boccaletti, M., Moratti, G., Gelati, R., Iaccarinof, S., Papani, G., Villa, G., 1999. Late Neogene evolution of the Taza–Guercif Basin (Rifian Corridor,Morocco) and implications for the Messinian salinity crisis. *Marine Geology*, **153**: 147–160.

Krijgsman, W., Blanc-Valleron, M.-M., Flecker, R., Hilgen, F. J., Kouwenhoven, T. J., Merle, D., Orszag-Sperber, F., Rouchy, J.-M., 2002. The onset of the Messinian salinity crisis in the eastern Mediterranean (Pissouri Basin, Cyprus). *Earth Planetary Science Letters*, **194**: 299– 300.

Krijgsman, W., Garcés, M., 2004. Paleomagnetic constraints on the geodynamic evolution of the Gibraltar Arc. *Terra Nova*, **16**: 281–287.

Krijgsman W., Meijer, P.T., 2008. Depositional environments of the Mediterranean “Lower Evaporites” of the Messinian Salinity Crisis: constraints from quantitative analyses. *Marine Geology*, **253**(3–4): 73–81.

Lewis, J., 1991. Cyst-theca relationships in Scrippsiella (Dinophyceae) and related orthoperidinoid genera. *Botanica marina*, **34**: 91-106.

Lewis, J., Harris, A.S.D., Jones, K.J., Edmonds, R.L., 1999. Long-term survival of marine planktonic diatoms and dinoflagellates in stored sediment samples. *Journal of Plankton Research*, **21**(2): 343-354.

Loget, N., Van Den Driessche, J., 2006. On the origin of the Strait of Gibraltar. *Sedimentary Geology*, **188-189**: 341-356.

Londeix, L., Benzakour, M., Suc, J.-P., Turon, J.-L., 2007. Messinian palaeoenvironments and hydrology in Sicily (Italy): The dinoflagellate cyst record Pale´oenvironnements et hydrologie du Messinien de Sicile (Italie) d'apre`s les kystes de dinoflagelle´s. *Geobios*, **40**: 233–250.

Lourens, L.J., Hilgen, F.J., Gudjonsson, L., Zachariasse, W.J., 1992. Late Pliocene to early Pleistocene astronomically forced surface productivity and temperature variations in the Mediterranean. *Marine Micropaleontology*, **19**: 49-78.

Lourens, L.J., Antonarakou, A., Hilgen, F.J., Van Hoof, A.A.M., Vergnaud-Grazzini,C., Zachariasse, W.J., 1996. Evaluation of the Plio-Pleistocene astronomical time- scale. *Paleoceanography*, **11**: 391–413.

Marret, F., Zonneveld, K., 2003. Atlas of modern organic-walled dino£agellate cyst distribution. *Revue of Palaeobotany and Palynology*, **125**: 1-200.

Martinez-Ruiz, F., Kastner, M., Paytan, A., Ortega-Huertas, M., Bernasconi, S.M., 2000. Geochemical evidence for enhanced produc- tivity during S1 sapropel deposition in the Eastern Mediterranean. *Paleoceanography*, **15**: 200–209.

Matano, F., Barbieri, M., Di Nocera, S., Torre, M., 2005. Stratigraphy and strontium geochemistry of Messinian evaporite-bearing successions of the southern Apennines foredeep, Italy: implications for the Mediterranean "salinity crisis" and regional Palaeogeography. *Palaeogeography, Palaeoclimatology, Palaeoecology*, **217**(1-2): 87-114.

Matthiessen, J., de Vernal, A., Head, M.J., Okolodkov, Y.B., Zonneveld, K.A.F., Harland, R., 2005. Modern organic-walled dinoflagellate cysts in Arctic marine environments and their (paleo) environmental significance. *Paläontologische Zeitschrift*, **79**: 3-51.

McCallum, J.E., Robertson, A.H.F., 1990. Pulsed uplift of the Troodos Massif: evidence from the Plio-Pleistocene Mesaoria Basin. In Malpas, J., Moores, E.M., Panayiotou, A., Xenophontos, C. (Eds.). Ophiolites: Oceanic Crustal Analogues. Proc. Symp. "Troodos 1987": *Department of Geological Survey, Ministry of Agriculture, Natural Resources and Environment*, Cyprus: 217–229.

McKenzie, J.A., Jenkyns, H., Bennett, G., 1979. Stable isotope study of the cyclic diatomite—clay-stones from the Tripoli Formation, Sicily: a prelude to the Messinian salinity crisis, *Palaeogeography, Palaeoclimatology, Palaeoecology*, **29**: 125–141.

Meier, K.J.S., Willems, H. 2003. Calcareous dinoflagellate cysts from surface sediments of the Mediterranean Sea: distribution patterns and influence of main environmental gradients. *Marine Micropaleontology*, **48**: 321-354.

Meier, K. J. S., Zonneveld, K. A. F., Kasten, S., Willems, H. 2004. Different nutrient sources forcing increased productivity during eastern Mediterranean S1 sapropel formation as reflected by calcareous dinoflagellate cysts. *Paleoceanography*, **19**. doi10.1029/2003PA000895: 1– 12.

Meier, K.J.S., Höll, C., Willems, H., 2004b. Effect of temperature on culture growth and cyst production in the calcareous dinoflagellates *Calciodinellum albatrosianum*, *Leonella granifera* and *Pernambugia tuberosa*. In: M. Triantaphyllou (Ed), Advances in the biology, ecology and taphonomy of extant calcareous nannoplankton. *Micropaleontology*, **50**(1): 93-106.

Meier, K.J.S., Young, J.R., Kirsch, M., Feist-Burkhardt, S., 2007. Evolution of different life-cycle strategies in the oceanic calcareous dinoflagellates. *European Journal of Phycology*, **42**(1): 81-89.

Meijer, P.T., Krijgsman, W., 2005. A quantitative analysis of the desiccation and re-filling of the Mediterranean during the Messinian Salinity Crisis. *Earth and Planetary Science Letters*, **240**(2): 510-520.

Meijer, P.T., 2006. A box model of the blocked-outflow scenario for the Messinian Salinity Crisis. *Earth and Planetary Science Letters*, **248**(1–2): 486–494.

Meyers, P.A., and Arnaboldi, M., 2005. Trans-Mediterranean comparison of geochemical paleoproductivity proxies in a Mid-Pleistocene interrupted sapropel. *Palaeogeography, Palaeoclimatology*, **222**: 313–328.

Merle, D., Lauriat-rage, A., Zorn, I., 2002. Les paléopeuplements marins du Messinien pré-évaporitique de Pissouri (Chypre , Méditerranée orientale) : aspects paléoécologiques précédant la crise de salinité messinienne. *Geodiversitas*, **24**: 669-689.

Meulenkamp, J., Dermitzakis, M., Georgiadou-Dikeoulia, E., Jonkers, H.A., Böger, H., 1979. Field guideto the Neogene of Crete. *Publications of the Department of Geology and Paleontology*, University of Athens, **A**: 1-15.

Montadert, L., Sancho, J., Fail, J.P., Debyser, J., Winnock, E., 1970. De l'age tertiare de la série salifére responsable des structures diapiriques en Méditerranée occidentale (Nord-est des Baleares). *Comptes rendus de l'Académie* des *sciences*, **271**: 812pp.

Montresor, M., Zingone, A., 1988. *Scrippsiella precaria* sp. nov. (Dinophyceae), a marine dinoflagellate from the Gulf of Naples. *Phycologia*, **3**: 387 394.

Montresor, M., Montesarchio, E., Marino, D., Zingone, A., 1994. Calcareous dinoflagellate cysts in marine sediments of the Gulf of Naples (Meditenanean Sea). Review *of* Palaeobotany *and Palynology*, **84**: 45-56.

Montresor, M., Janofske, D., Willems, H., 1997. The cyst-theca relationship in *Calciodinellum operosum emend.* (Peridiniales, Dinophyceae) and a new approach for the study of calcareous cysts. *Journal of Phycolology*, **33**: 122-131.

Montresor, M., Zingone, A. and Sarno, D., 1998. Dinoflagellate cyst production at a coastal Mediterranean site. *Journal of Plankton Research*, **20**(12): 2291-2312.

Moores,E.M., Vine, F.J., 1971. The Troodos Massif, Cyprus and other ophiolites as oceanic crust: evaluation and implications. *Philosophical Transactions of the Royal Society of London*, **268**(A): 433-466.

Morris, A., Tarling, D.H., 1996. Palaeomagnetism and tectonics of theMediterranean region: an introduction. In Morris, A., Tarling, D.H., (Eds.). Palaeomagnetism and tectonics of the Mediterranean region. *Geological Society Special Publication,* **105**:1–18.

Orszag-Sperber, F., Rouchy, J.M., Elion, P., 1989. The sedimentary expression of regional tectonic events during the Miocene-Pliocene transition in the southern Cyprus basins. *Geological Magazine*, **126**(3): 291-299.

Orszag-Sperber, F, Rouchy, J.-M., Blanc-Valleron, M.-M., 2000. La transition Messinien–Pliocène en Méditerranée orientale (Chypre): la période du *Lago-Mare* et sa signification. *Earth and Planetary Sciences*, **331**: 483-490.

Orszag-sperber, F., 2006. Changing perspectives in the concept of "Lago-Mare" in Mediterranean Late Miocene evolution. *Sedimentary Geology*: doi:10.1016/j.sedgeo.2006.03.008.

Orszag-Sperber, F., Caruso, A., Blanc-Valleron, M.-M., Merle, D., Rouchy, J.M., 2009. The onset of the Messinian salinity crisis: Insights from Cyprus sections. *Sedimentary Geology*, **217**: 52-64.

Pearce, J.A., Bender, J.F., De Long, S.E., Kidd, W.S.F., Low, P.J., Guner, Y., Saröglü, F., Yilmaz, Y., Moorbath, S., Mitchell, J.G., 1990. Genesis of collisional volcanism in eastern Anatolia, Turkey. *Journal of Volcanology Geothermal Research*, **44**: 189–229.

Parsons, T.R., Takahashi, M., Hargrave, B., 1984. Biological oceanographic processes. Pergamon Press, Oxford: 330pp.

Pedley, H.M., Grasso, M., (1993). Controls on faunal and sediment cyclicity within the Tripoli and Calcare di Base basins (Late Miocene) of central Sicily. *Palaeogeography, Palaeoclimatology, Palaeoecology*, **105**: 337-360.

Pierre, C., Rouchy, J.-M., Blanc-Valleron, M.-M., 1998. Sedimentological and stable isotope changes at the Messinian/Pliocene boundary in the Eastern Mediterranean (Holes 968A, 969A, and 969B). *Proceedings of the Ocean Drilling Program, Scientific Results*, 160: 3-8.

Pinet, P.R., 1998. Invitation to Oceanography. *Jones and Bartlett Publishers*: 508pp.

Richter, D., Vink, A., Zonneveld, K.A.F., Kuhlmann, H., Willems, H., 2007. Calcareous dinoflagellate cyst distributions in surface sediments from upwelling areas off NW Africa, and their relationships with environmental parameters of the upper water column. *Marine Micropaleontology* **63**(3–4): 201–228.

Robertson, A.H.F., Hudson, J.D., 1974. Pelagic sediments in the Cretaceous and Tertiary history of the Troodos Massif, Cyprus. In: Pelagic Sediments: On Land and Under the Sea Hsü, K.J.; Jenkyns, H.C. (Eds.). International *Association of Sedimentology, Spececial Publication*, **1**: 403–436.

Rizzini, A., Vezzani, F., Cocoecetta, V. and Milad, G., 1978. Stratigraphy and sedimenta- tion of a Neogene--Quaternary section in the Nile Delta area (A.R.E.). *Marine Geology*, **27**: 327-348.

Robertson, A.H.F., Eaton, S.E., Folloes, E.J., Payne, A.S., 1995. Sedimentology and depositional processes of Miocene evaporites from Cyprus. *Terra Nova*, **7**: 233-254.

Robertson, A.H.F., Grasso, M., 1995. Overview of the Late Tertiary– Recent tectonic and paleo-environmental development of the Medi- terranean region. *Terra Nova*, **7**: 114–127.

Robertson, A., 1998. Tectonic significance of the Eratosthenes Seamount: a continental fragment in the process of collision with a subduction zone in the eastern Mediterranean (Ocean Drilling Program Leg 160). *Tectonophysics*, **298**: 63-82.

Rochon, A., 2009. The ecology and biological affinity of Arctic dinoflagellates and their paleoceanographical significance in the Canadian High Arctic. *IOP Conference Series: Earth and Environmental Science*, **5**: 12003.

Rögl, F., 1999. Mediterranean and Paratethys; facts and hypotheses of an Oligocene to Miocene paleogeography; short overview. *Geologica Carpathica*, **50**(4): 339–349.

Rohling, E.J., Hilgen, F.J., 1991. The eastern Mediterranean climate at times of sapropel formation: a review. *Geologie en Mijnbouw:* 253–264.

Rohling, E.J., 1994. Review and new aspects concerning the formation of eastern Mediterranean sapropels. *Marine Geology*, **122**: 1–28.

Rossignol-Strick , M., 1983. African monsoons, an immediate climate response to orbital insolation. *Nature*, **304**: 46–49.

Rossignol-Strick, M. 1985. Mediterranean Quaternary sapropels, an immediate response of the African monsoon to variation of insolation. *Palaeogeography, Palaeoclimatology, Palaeoecology*, **49**: 237-263.

Rouchy, J.M., 1982. La genèse des évaporites messiniennes de Méditerranée. *Mémoires du Muséum National d'Histoire Naturelle*, Paris **50**(C): 267pp.

Rouchy, J.M., Saint-Martin, J.P., 1992. Late Miocene events in the Mediterranean as recorded by carbonate–evaporite relations. *Geology*, **20**: 629–632

Rouchy, J.M., Orszag-Sperber, F., Blanc-Valleron, M.-M., Pierre, C., Rivière, M., Combourieu-Nebout, N., Panayides, I. 2001. Paleoenvironmental Changes at the Messinian–Pliocene Boundary in the Eastern Mediterranean (Southern Cyprus basins): Significance of the Messinian Lago –Mare. *Sedimentological Geology*, **145**: 93–117.

Rouchy, J.M., Caruso, A., 2006. The Messinian Salinity Crisis in the Mediterranean basin: a reassessment of the data and an integrated scenario. *Sedimentary Geology*, **188–189**: 35–67.

Roveri, M., Bassetti, M.A., Ricci Lucchi, F., 2001. The Mediterranean Messinian salinity crisis: an Apennine foredeep perspective. *Sedimentary Geology*, **140** (3–4): 201–214.

Roveri, M., Scienze, V., Zamboni, V., Rogledi, S., 2003. Sedimentary and tectonic evolution of the Vena del Gesso basin (Northern Apennines , Italy): Implications for the onset of the Messinian salinity crisis. *America*: 387-405.

Roveri, M., Taviani, M., 2003. Calcarenite and sapropel deposition in the Mediterranean Pliocene: shallow- and deep-water record of astronomically driven climatic events. *Terra Nova*, **15**: 279-286.

Roveri, M., Manzi, V., 2006. The Messinian Salinity Crisis: looking for a new paradigm? *Palaeogeography, Palaeoclimatology, Palaeoecology*, **238**: 386–398.

Roveri, M., Bertini, A., Cipollari, P., Cosentino, D., Di, Stefano, A., Florindo, F., Gennari, R., Gliozzi, E., Grossi, F., Iaccarino, S. Luigli, S., Manzi, V., 2008. Earliest Zanclean age for the Colombacci and uppermost Di Tetto formations of the latest Messinian northern Apennines: New palaeoenvironmental data from the Maccarone section (Marche Province, Italy) by Popescu et al. (2007) Geobios 40 (359-373). *Geobios*, **41**: 669-675.

Ruggieri, G., Sprovieri, R., 1976. Messinian salinity crisis and its paleogeographical implications. *Palaeogeography, Palaeoclimatology, Palaeoecology*, **20**: 13–21.

Ryan, W.B.F., Cita, M.B., 1978. The nature and distribution of the Messininan erosional surface—indicators of a several-kilometers- deep Mediterranean in the Miocene. *Marine Geology*, **27**: 193–230.

Ryan, W.B.F., 2009. Decoding the Mediterranean salinity crisis. *Atlantic*: 95-136.

Sanders, R.W., 1991. Mixotrophic protists in marine and freshwater ecosystems. *Journal of Protozoology*, **38**: 76–81.

Saldarriaga, J.F., Taylor, F.J.R., Keeling, P.J., Cavalier-smith, T., 2001. Dinoflagellate Nuclear SSU rRNA Phylogeny Suggests Multiple Plastid Losses and Replacements. *Journal of Molecular Evolution*, **53**:204–213.

Schnepf, E., Elbrächter, M., 1992. Nutritional strategies in dinoflagellates: a review with emphasis on cell biological aspects. *Journal of Protisto1ology*, **28**: 3-24.

Selli, R., 1960. Il Messiniano Mayer-Eymar 1867. Proposta di un neostratotipo. *Giornale di Geologia*: **28**: 1–33.

Sgarrella, F., Sprovieri, R., Di Stefano, E., Caruso, A., 1997. Paleoceanographic conditions at the base of the Pliocene in the Southern Mediterranean Basin. *Rivista Italiana di Paleontologia i Stratigrafia*, **103**: 207–220.

Siano, R., Montresor, M., 2005. Morphology, ultrastructure and feeding behaviour of Protoperidinium vorax sp. nov. (Dinophyceae, Peridiniales). *European Journal of Phycology*, **40**: 221-232.

Sierro, F.J., Hilgen, F.J., Krijgsman, W., Flores, J.A., 2001. The Abad composite (SE Spain): a Messinian reference section for the Mediterranean and the APTS. *Atlantic*, **168**: 141-169.

Sikes, C.S., Wierzbicki, A., Fabry, V.J., 1994. From atomic to global scales in biomineralization. *Bulletin de l'Institut Oceanographique*, **14**(1): 1-47.

Smayda, T.J., 2000. Ecological features of harmful algal blooms in coastal upwelling ecosystems. *South African Journal of Marine Science*, **22**: 219–253.

Smayda, T.J., 2002. Turbulance watermass stratification and harmful algal blooms: an alternative view and frontal zones as pelagic seed banks. *Harmful Algae*, **1**: 95–112.

Smayda, T.J., 2010. Adaptations and selection of harmful and other dinoflagellate species in upwelling systems. 2. Motility and migratory behaviour. *Progress In Oceanography*, **85**: 71-91.

Smayda, T.J., Reynolds, C.S., 2001. Community assembly in marine phytoplankton: application of recent models to harmful dinoflagellate blooms. Journal of Plankton Research, **23**: 447–61.

Smayda, T.J. ,Trainer, V.L., 2010. Dinoflagellate blooms in upwelling systems: Seeding, variability, and contrasts with diatom bloom behaviour. *Progress In Oceanography*, **85**: 92-107.

Sprovieri, R., Di Stefano, E., Sprovieri, M., 1996a. High resolution chronology for late Miocene Mediterranean stratigraphic events. *Rivista Italiana di Paleontologia e Stratigrafia*, **102**: 77–104.

Sprovieri, R., Di Stefano, E., Caruso, A., Bonomo, S., 1996b. High resolution stratigraphy in the Messinian Tripoli Formation in Sicily. *Paleopelagos*, **6**: 415–435.

Stampfli, G.M., Höcker, C.F.W., 1989. Messinian paleorelief from a 3-D seismic survey in the Tarraco concession area (Spanish Mediterranean sea). *Geologie en Mijnbouw*, **68**(2): 201–210.

Stampfli, G.M., Borel, G., 2004. The TRANSMED transects in space and time: constraints on the paleotectonic evolution of the Mediterranean domain. In: Cavazza, W., Roure, F.M., Spakman, W., Stampfli, G.M., Ziegler, P.A. (Eds.). The TRANSMED Atlas. The Mediterranean Region from Crust to Mantle. *SpringerVerlag*, Berlin–Heidelberg: 53–90.

Stickney, H., 2000. The impact of mixotrophy on planktonic marine ecosystems. *Ecological Modelling*, **125**: 203-230.

Stoecker, D.K., 1999. Mixotrophy among dinoflagellates. *Journal of Eukaryotic Microbiology*, **46**: 397–401.

Stover, L.E., Brinkhuis, H., Damassa, S.P., de Verteuil, L., Helby, R.J., Monteil, E., Partridge, A., Powell, A.J., Riding, J.B., Smelror, M., Williams, G.L., 1996. Mesozoic–Tertiary dinoflagellates, acritarchs and prasinophytes. In: Jansonius, J., McGregor, D.C. (Eds.). Palynology: Principles and Applications. *American Association of Stratigraphic Palynologists Foundation*: 641–750.

Stow, D.A.V., Braakenburg, N.E., Xenophontos, C., 1995. The Pissouri Basin fan-delta complex, southwestern Cyprus. *Sedimentary Geology*, **98**: 254-262.

Streng, M., Hildebrand-Habel, T., Willems, H., 2004. Long-term evolution of calcareous dinoflagellate associations since the Late Cretaceous: comparison of a high- and a low-latitude core from the Indian Ocean. *Journal of Nannoplankton Research*, **26**: 13-45.

Streng, M., Hildebrand-Habel, T. ,Willems, H., 2004. a Proposed Classification of Archeopyle Types in Calcareous Dinoflagellate Cysts. *Journal of Paleontology*, **78**: 456-483.

Suc, J.-P., Bessais, E., 1990. Pérennité d'un climat thermoxérique en Sicile, avant, pendant et après la crise de salinité messinienne. *Compte Rendus Academy of Science*, **310**(II): 1701–1707.

Suc, J.-P., Cravatte, J., 1982. Etude palynologique du Pliocène de Catalogne (nord-est de l'Espagne): apports à la connaissance de l'histoire climatique de la Méditerranée occidentale et implications chronostratigraphiques. *Paléobiologie Continentale*, **13**(1): 1–31.

Suc, J.-P., Violanti, D., Londeix, L., Poumot, C., Robert, C., s Clauzon, G., s Gautier, F., Turon, J.-L., Ferrier, J. Chikhi, H., Cambon, G., 1995. Evolution of the Messinian Mediterranean environments: the Tripoli Formation at Capodarso (Sicily, Italy). *Review of Palaeobotany and Palynology*, **87**(1): 51-79.

Sullivan, J.M., Swift, E., 2003. Effects of small-scale turbulence on the net growth rate and size of ten species of marine dinoflagellates. *Journal of Phycology*, **39**(1): 83-94.

Sullivan, C.W., Arrigo, K.R., McClain, C.R., Cosimo, J.C., Firestone, J., 1993. Distribution of phytoplankton blooms in the Southern Ocean. *Science*, **262**: 1832–1837.

Tangen, K., Brand, L.E., Blackwelder, P.L. and Guillard, R.R.L., 1982. Thoracosphaera heimii (Lohmann) Kamptner is a dinophyte: Observations on its morphology and life cycle. *Marine Micropaleontology*, **7**: 193-212.

Taylor, F.J.R., 1987. General and marine environments. In: Taylor, F.J.R. (Ed.). The Biology of Dinoflagellates. *Botanical Monographs, Blackwell Scientific Publications*, **21**: 399-502.

Taylor, F.J.R. and Pollingher, U., 1987. The ecology of dinoflagellates. In: Taylor, F.J.R. (Ed.). The biology of dinoflagellates. *Blackwell Scientific Publications*: 398-529.

Taylor, F., Hoppenrath, M., Saldarriaga, J., 2008. Dinoflagellate diversity and distribution. *Biodiversity and Conservation*, **17**(2): 407-418.

Thunell, R.C., Williams, D.F., Belyea, P.R., 1984. Anoxic events in the Mediterranean Sea in relation to the evolution of late Neogene climates. *Marine* Geology, **59**: 105–134.

Van Assen, E., Kuiper, K.F., Barhoun, N., Krijgsman, W., Sierro, F.J., 2006. Messinian astrochronology of the Melilla Basin: stepwise restriction of the Mediterranean–Atlantic connection through Morocco. *Palaeogeography, Palaeoclimatology, Palaeoecology*, **238**: 15–31.

Van der Laan, E., 2005. Regional climate and glacial control on high-resolution oxygen isotope records from Ain el Beida (latest Miocene, northwest Morocco): A cyclostratigraphic analysis in the depth and time domain. *Paleoceanography*, **20**: 1-22.

Van Couvering, G.J., Berggren, W.A., Drake, R.E., Aguirre, E., Curtis, G.H., 1976. The terminal Miocene event. *Marine Micropaleontology*, **10**: 71–90.

Van der Laan, E., Snel, E., de Kaenel, E., Hilgen, F. J., Krijgsman, W. 2006. No major deglaciation across the Miocene-Pliocene boundary: Integrated stratigraphy and astronomical tuning of the Loulja sections (Bou Regreg area, NW Morocco). *Paleoceanography*, **21**: 1-27.

Van der Zwaan, G.J., Gudjonsson, L., 1986. The Middle Miocene–Pliocene stable isotope stratigraphy and paleoceanography of the Mediterranean. *Marine Micropaleontology*, **10**: 71–90.

Van Dijk, J.P., Barberis, A., Cantarella, G., Massa, E., 1998. Central Mediterranean Messinian basin evolution. Tectono-eustasy or eustato-tectonics? *Annales Tectonicae*, **12**(1-2): 7-27.

Vazquez, A., Utrilla, R., Zamarrenos, G., Sierro, F.J., Flores, J.A., Barcena, M.A., 2000. Erlenkeuser, H., Barcena, M.A., 2000. Precession related sapropelites of the Messinian Sorbas Basin (South Spain): Paleoenvironmental significance. *Palaeogeography, Palaeoclimatology, Palaeoecology*, **158**: 353-370.

Vergnaud-Grazzini, C., 1983. Reconstruction of Mediterranean Late Cenozoic hydrography by means of carbon isotope analyses. *Micropaleontological Bulletin*, Utrecht, **30**: 25–47.

Versteegh, G.J.M., 1993. New Pliocene and Pleistocene calcareous dinoflagellate cysts from southern Italy and Crete. *Review of Palaeobotany and Palynology*, **78**: 353-380.

Vink, A., 2004. Calcareous dinoflagellate cysts in South and equatorial Atlantic surface sediments: diversity, distribution, ecology and potential for palaeoenvironmental reconstruction. *Marine Micropaleontology*, **50**: 43-88.

Vink, A., Brune, A., Höll, C., Zonneveld, K.A.F., Willems, H., 2002. On the response of calcareous dinoflagellates to oligotrophy and stratification of the upper water column in the equatorial Atlantic Ocean. *Palaeogeography, Palaeoclimatology, Palaeoecology*, **178**: 53-66.

Vink, A., Rühlemann, C., Zonneveld, K.A.F., Mulitza, S., Hüls, M., Willems, H., 2001. Shifts in the position of the North Equatorial Current and rapid productivity changes in the western Tropical Atlantic during the last glacial. *Paleoceanography*, **16**(5): 479-490.

Vink, A., Zonneveld, K.A.F., Willems, H., 2000. Distributions of calcareous dinoflagellates in surface sediments of the western equatorial Atlantic, and their potential use in palaeoceanography. *Marine Micropaleontology*, **38**: 149-180.

Wade, B.S., Bown, P.R., 2005. Calcareous nannofossils in extreme environments: The Messinian Salinity Crisis, Polemi Basin, Cyprus. *Palaeogeography, Palaeoclimatology, Palaeoecology*, **233**(3-4): 271-286.

Wall, D., Dale, B., 1968. Quaternary calcareous dinoflagellates (Calciodinelloideae) and their natural affinities. *Journal of Paleontol*ogy, **42**: 1395-1408.

Wall, D., Guillard, R.R.L., Dale, B., Swift, E. and Watabe, N., 1970. Calcitic resting cysts in Peridinium trochoideum (Stein) Lemmermann, an autotrophic marine dinoflagellate. *Phycologia*, **9**(2): 151-156.

Wendler, J., Wendler, I., Willems, H., 2001. Orthopithonella collaris sp. nov., a new calcareous dinoflagellate cyst from the K/T boundary (Fish Clay, Stevns Klint/Denmark). *Review of Palaeobotany Palynology*, **115**: 69-77.

Wendler, J., Willems, H., 2002. The distribution of pattern of calcareous dinoflagellate cysts at the K/T boundary (Fish Clay, Stevns Klint, Denmark) - Implications for our understanding of species selective extinction. In: Koeberl, C., MacLeod, K.G. (Eds.). Catastrophic events and mass extinctions: impacts and beyond. *Geological Society of America Special Paper*, **356**: 265-276.

Wendler, I., Zonneveld, K. A. F., Willems, H., 2002a. Calcareous cyst-producing dinoflagellates: ecology and aspects of cyst preservation in a highly productive oceanic region. *In*: Clift, P.D., Kroon, D., Geadicke, C., Craig, J. (Eds.). The tectonic and climatic evolution of the Arabian Sea region. *Geological Society Special Publication*, **195**: 317-340.

Wendler, I., Zonneveld, K. A. F., Willems, H., 2002b. Oxygen availability effects on early diagenetic calcite dissolution in the Arabian Sea as inferred from calcareous dinoflagellate cysts. *Global and Planetary Change*, **34**: 219-239.

Wendler, I., Zonneveld, K. A. F., Willems, H., 2002c. Production of calcareous dinoflagellate cysts in response to monsoon forcing off Somalia: a sediment trap study. *Marine Micropaleontology*, **46**: 1-11.

Willems, H., 1988. Kalkige Dinoflagellaten-Zysten aus der oberkretazischen Schreibkreide-Fazies N-Deutschlands (Coniac bis Maastricht). *Senckenbergiana lethaea*, **68**(5): 433-477.

Willems, H., 1996. Calcareous dinocysts from the Geulhemmerberg K/T boundary section (Limburg, SE Netherlands). *Geologie Mijnb.*, **75**: 215-231.

Williams, D.F., Thunell, R.C. and Kennett, J.P., 1978. Periodic freshwater flooding and stagnation of the eastern Mediterranean Sea during the late Quaternary. *Science*, **201**: 252-254.

Young, J. R., Bergen, J. A, Bown, P. R., Burnett, J. A, Fiorentino, A, Jordan, R. W., Kleijne, A, van Niel, B. E., Romein, A. J. T., von Salis, K., 1997. Guidelines for coccolith and calcareous nannofossil terminology. *Palaeontology*, **40**(4): 875-912.

Zonneveld, K.A.F., Höll, C., Janofske, D., Karwath, B., Kerntopf, B., Rühlemann, C. , Willems, H., 1999. Calcareous dinoflagellates as palaeo-environmental tools: 145-164p. *In* Fischer, G., Wefer, G. (Eds.), *Use of proxies in Paleoceanography: Examples from the South Atlantic*. *Springer Verlag*, Berlin: 735 pp.

Zonneveld, K.A. F., Brune, A.,Willems, H., 2000. Spatial distribution of calcareous dinoflagellates in surface sediments of the South Atlantic Ocean between 13°N and 36°S. *Review of Palaeobotany and Palynology*, **111**: 197-223.

Zonneveld, K.A.F., Versteegh, G.J.M., De Lange, G. J., 2001. Palaeoproductivity and post-depositional aerobic organic matter decay reflected by dinoflagellate cyst assemblages of the Eastern Mediterranean S1 sapropel. *Marine Geology*, **172**: 181-195.

Zonneveld, K.A.F., 2004. Potential use of stable oxygen isotope composition of *Thoracosphaera heimii* for upper water column (thermocline) temperature reconstruction. *Marine Micropaleontology*, **50**(3/4): 307-317.

Zonneveld, K.A.F., Meier, K.J.S., Esper, O., Siggelkow, D., Wendler, I., Willems, H., 2005. The (palaeo-)environmental significance of modern calcareous dinoflagellate cysts: a review. *Paläontologische Zeitschrift*, **79**(1): 61–77.

Zügel, P., 1994. Verbreitung kalkiger Dinoflagellaten-Zysten im Cenoman/Turon von Westfrankreich und Norddeutschland. *Courier Forschungs Institut Senckenberg*, **176**: 159pp.

Chapter 2

Calcareous dinoflagellate turnover in relation to the Messinian salinity crisis in the eastern Mediterranean Pissouri Basin, Cyprus

Katarzyna-Maria Bison[1*], Gerard J.M. Versteegh[2], Frits J. Hilgen[3] & Helmut Willems[1]

[1]*Division of Palaeontology, University of Bremen, FB 5, Geowissenschaften, Postfach 330440, D-28334 Bremen, Germany*

[2]*Institut für Biogeochemie und Meereschemie, Universität Hamburg, Bundesstraße 55, D-20146 Hamburg, Germany.*

[3]*University of Utrecht, Faculty of Earth Sciences, Budapestlaan 4, 3594 Utrecht, Netherlands.*

[*]Corresponding author: kbison@uni-bremen.de

***Journal of Micropalaeontology* (2007), 26: 103–116.**

(Manuscript received 26 July 2005 Manuscript accepted 23 January 2007)

Abstract

The extent to which the Messinian salinity crisis modified the initially Tethyan, eastern Mediterranean phytoplankton community has been investigated by monitoring the fate of calcareous dinoflagellate cyst assemblages prior to, during and after the Salinity Crisis in the Pissouri Section (Cyprus). A rich, but low diversity open oceanic assemblage, dominated by *Calciodinellum albatrosianum,* occurs during the upper Tortonian and lower Messinian. The upper Messinian (pre-evaporitic) sediments yield only few cysts but the assemblage is much more diverse and reflects unstable more neritic conditions (*Bicarinellum tricarinelloides*), fluvial influence (*Leonella granifera*) and varying, temporally increased salinities (*Pernambugia tuberosa*), probably related to the increasingly restricted environment. The basal Pliocene sediments reflect the return to normal marine conditions; the dinoflagellate assemblage is rich in cysts and again has a low diversity. However, in contrast to the *C. albatrosianum*-dominated upper Tortonian and pre-evaporitic Messinian sediments, *L. granifera* clearly dominates the basal Pliocene association just after the replenishment of the Mediterranean basin. Apart from this shift in dominance, the onset of the Pliocene is furthermore marked by the first appearance of *Calciodinellum elongatum*, which must have immigrated from the Atlantic Ocean. *Lebessphaera urania*, a postulated remnant of the Tethyan Ocean survived the salinity crisis, possibly in as yet unidentified marine refuges in the Mediterranean itself. Although the environmental changes caused by the Messinian salinity crisis did not lead to an extinction of calcareous dinoflagellate species of the Pissouri Basin, it resulted in a significant change in the assemblages and contributed to a more modern character of the Pliocene dinoflagellate association in the Eastern Mediterranean.

Keywords: Messinian, calcareous dinoflagellates, Mediterranean, Pliocene, Miocene

Introduction

The dramatic drop in the sea level of the Mediterranean Basin during the Messinian salinity crisis represents a major environmental breakdown within an oceanic basin. It led to extensive and widespread evaporite deposition underlying the modern Mediterranean Sea (e.g. Hsü, 1973, 1978; Hsü et al., 1977; Butler *et al.*, 1999; Lofi *et al.*, 2005; Meijer & Krijgsman, 2005; Tay *et al.*, 2002). These evaporites reflect the progressive closure of the Mediterranean gateways, which resulted in the almost complete isolation of the Mediterranean from the Atlantic Ocean during the final stage of the Messinian salinity crisis (Hsü & Bernoulli, 1978; Krijgsman *et al.*, 1999; Bianchi & Morri 2000; Seidenkrantz *et al.*, 2000; Blanc, 2002; Manzi *et al.*, 2005; Matano *et al.*, 2005).

Considering that the Mediterranean Sea is an old Tethys remnant, it is a natural heir to the Tethyan biodiversity (Bianchi & Morri, 2000; Boullon *et al.*, 2004). Despite the connection between the Atlantic and the Mediterranean, one may expect that migration of Atlantic species into the Mediterranean must have been largely restricted to those niches that were not already occupied by the Tethys biota. The Messinian Salinity Crisis may have changed this. The drastic environmental changes during the Salinity Crisis caused by the significant fall in sea level (e.g. Loget *et al.*, 2006) must have wiped out, or diminished many taxa (Bouchet & Taviani, 1992; Bianchi & Morri, 2000; Seidenkrantz *et al.*, 2000; Logan *et al.*, 2004; Domingues *et al.*, 2005; Kouwenhoven *et al.*, 2006). In the Mediterranean, the relatively stable open oceanic environment probably disappeared completely, in contrast to the neritic realms, which in part experienced a displacement only (Keogh & Butler, 1999; Bianchi & Morri, 2000). Therefore, oceanic extinction must have been most severe (Bianchi & Morri, 2000; Logan *et al.*, 2004), especially for deep-water representatives of most phyla, a remarkable decrease was recognised (Seidenkrantz *et al.*, 2000; Emig & Geistdoerfer, 2004; Kouwenhoven *et al.*, 2006). After the crisis, the vacant niches will have been refilled by species migrating or re-migrating from the Atlantic into the Mediterranean (Bianchi & Morri, 2000; Logan *et al.*, 2004). An important biogeographical question is attached to this event: how and to what extent did the originally Tethyan marine biota survive this crisis and to what extent were they replaced by newcomers from the adjacent Atlantic?

Calcareous dinoflagellates could provide important insight in the development of the Messinian environmental change and its biogeographical effects since they react sensitively to salinity, nutrient and temperature changes and they have representatives in the entire spectrum of marine environments, from open marine to estuarine (e.g. Dale, 1992; Höll *et al.*, 1998, 1999; Zonneveld *et al.*, 1999; Höll & Kemle von Mücke, 2000; Wendler *et al.*, 2002a, b, c; Meier & Willems, 2003; Tanimura & Shimada, 2004; Vink, 2004; Kohring *et al.*, 2005).

Unfortunately, information on Miocene and Pliocene calcareous dinoflagellate assemblages from this region is very sparse. Comparison of the associations from the few investigated localities from the Mediterranean area (Keupp & Kohring, 1993, 1999; Kohring, 1993b, 1997; Keupp *et al.*, 1994) shows strong changes in cyst abundance but only little variability in species diversity and composition. On the basis hereof, an alternative scenario to that postulated above has been proposed: thereby a permanent input of oceanic species from the Atlantic into the Mediterranean during the Miocene – Pliocene which kept the oceanic assemblages of the Atlantic and Mediterranean more or less identical (Keupp & Kohring,1993, 1999; Kohring, 1993b). In this scenario, the Messinian Salinity Crisis is supposed to have affected the neritic species only. To shed more light on this controversy, the calcareous dinoflagellate assemblages from the eastern Mediterranean upper Miocene-lower Pliocene sedimentary succession from the Pissouri Basin on Cyprus were investigated.

Modern calcareous dinoflagellates

Calcareous dinoflagellates are primary producers and thus are restricted to the euphotic zone (e.g. Höll *et al.*, 1999; Wendler *et al.*, 2002b; Vink *et al.*, 2002; Vink, 2004). In modern oceans, they often dominate the total dinoflagellate flux to the sea floor (Dale, 1992; Zonneveld *et al.*, 1999; Vink *et al.*, 2000). In the Mediterranean, calcareous dinoflagellates are the second most important carbonate producers after coccolithophorids with an average of 17 % of the total biogenic carbonate flux (Ziveri *et al.*, 2000). Over the last decades, our understanding of their phylogeny, biodiversity and ecology has greatly increased (e.g. Keupp & Versteegh, 1989; Höll *et al.*, 1998, 1999; Zonneveld *et al.*, 1999, 2000; Karwath *et al.*, 2000; Wendler *et al.*, 2002b, c; Friedrich & Meier, 2003; Gottschling & Plötner, 2004; Meier *et al.*, 2004; Tanimura & Shimada, 2004; Vink, 2004; Zonneveld, 2004). This knowledge has been applied successfully to reconstruct environmental parameters such as productivity and surface water stratification (Höll *et al.*, 1999; Höll & Kemle-von Mücke, 2000; Esper *et al.*, 2000; Wendler *et al.*, 2002c; Vink, 2004; Meier *et al.*, 2004).

Mediterranean assemblages

The Mediterranean associations have been shown to differ from those of other marine environments. The apparent gradual west–east change in the Mediterranean dinoflagellate cyst assemblages is strongly correlated with the main environmental gradients in the surface waters (Meier & Willems, 2003; Meier *et al.*, 2004). Eastwards, salinity and temperature increase whereas nutrient concentrations decrease.

Today, *Thoracosphaera heimii* dominates the Mediterranean assemblages with relative abundances >88%, whereby its relative abundance in the Eastern Basin is usually lower than in the Western Basin (Meier & Willems, 2003). *T. heimii* also dominates the Arabian Sea, Atlantic and Pacific Oceans but if we exclude *T. heimii,* the most abundant species are *Calciodinellum albatrosianum*, *Leonella granifera*, *C. levantinum* and *Pernambugia tuberosa* (e.g. Wendler *et al.*, 2002a; Hernández-Becerril & Bravo-Sierra, 2004; Tanimura & Shimada, 2004; Vink, 2004). From these taxa only *C. levantinum* reaches similar high relative abundances in the western Mediterranean Sea (Meier & Willems, 2003) which is considered to reflect an Atlantic influence through the Strait of Gibraltar (Meier & Willems, 2003). This species is closely similar to, and has been confused with *P. tuberosa*, which is one of the main species in the tropical and South Atlantic Ocean (Höll *et al.*, 1999; Esper *et al.*, 2000).

Lebessphaera urania is only known from the eastern Mediterranean where it dominates the dinoflagellate associations (Meier *et al.*, 2002; Meier & Willems, 2003). *Scripsiella regalis* is a frequent component in the whole Mediterranean Sea. *C. albatrosianum* represent <5 % of the associations only, whereas *Pernambugia tuberosa* (former name *Sphaerodinella tuberosa*) is absent (Meier & Willems, 2003) but forms up to 75% in the South and equatorial Atlantic Ocean (Vink, 2004).

L. granifera is mostly rare to absent in Mediterranean sediments or traps (Meier & Willems, 2003) but it reaches high relative and absolute abundances in more shelfward regions of the Atlantic Ocean and Arabian Sea where river influence and / or continental nutrient supply (dust) play a major role (Vink, 2000, 2004; Zonneveld *et al.*, 2001; Wendler *et al.*, 2002b, c; Meier *et al.*, 2004; Tanimura & Shimada, 2004). This species appears to be an excellent proxy for nutrient enriched environments induced by terrigenous input (e.g. Vink, 2004; Meier *et al.* 2004; Wendler *et al.*, 2002b, c).

This interpretation of *L. granifera* is applied to re-interpret environmental reconstructions of other Miocene and Pliocene assemblages, leading to a better understanding of past environmental settings. Until now, the fossil distribution pattern of *L. granifera* could not be related to any environmental parameter except as a general open marine species with no distinct ecological preferences (Fütterer, 1977; Keupp & Kohring, 1993; Kohring 1993, 1997).

Palaeoenvironment of the Pissouri Basin

The Neogene sedimentary succession on Cyprus shows a gradual shallowing from Palaeogene deep-water pelagic carbonates (Lefkara Formation) through mixed detrital sediments, carbonates and reefal limestones, to the Messinian evaporites (Orszag-Sperber *et al.*, 1989; Robertson *et al.*, 1995; Krijgsman *et al.*, 2002). The Messinian evaporites on Cyprus have been deposited in small sub-basins (e.g. Polemi, Psematismenos and Pissouri Basins) (Robertson *et al.*, 1995; Orszag-Sperber, 1989; Krijgsman *et al.*, 2002). With the beginning of the Pliocene, deeper water conditions were re-established (Rouchy *et al.*, 2001; Krijgsman *et al.*, 2002). The Messinian sedimentary succession on Cyprus is characterized by pre-evaporitic deposits, overlain by two gypsum units followed by a lagoonal-lacustrine transitional facies (Lago Mare facies) leading to the marine marls of the basal Pliocene (Krijgsman *et al.*, 2002; Rouchy *et al.* 2001). A small hiatus between 5.59 and 5.52 Ma (Carnevale *et al.* 2006) representing an erosive phase during the Late Messinian lowstand occurs between the upper gypsum deposits and the overlaying lowermost Pliocene sediments (Orszag-Sperber *et al.* 2000).

Palaeodepth estimations in the Pissouri Section are based on foraminifera, bivalves and gastropods (Rouchy *et al.* 2001; Krijgsman *et al.*, 2002; Kouwenhoven *et al.* 2006). A progressive shallowing occurred from approximately 500 m during the Tortonian and Early Messinian to very shallow water conditions during the late Messinian. Sediments above Pissouri Cycle 9 (Fig. 2) were probably deposited at depths less than 100 m (Krijgsman *et al.*, 2002). Gypsum and its pseudomorphs already occur sporadically above cycle 22 (Fig. 2) with an increasing frequency and amount toward the evaporitic stage of the salinity crisis (Krijgsman *et al.*, 2002). For the earliest Pliocene of the Pissouri Basin a depth of at least 300 m has been estimated with a deepening trend (Rouchy *et al.*, 2001).

Studies on Neogene and Pleistocene sediments from the Gulf of Suez and the Red Sea and on drilling sites of the Ocean Drilling Programm (ODP-sites 658-661) resulted in the assumption that northeastern Africa had a relatively humid climate during the late Messinian and was strongly influenced by high monsoonal activity (Griffin, 1999). This precipitation-rich period peaked into the late Messinian during the low-stand in the Mediterranean Basin and led to increased sedimentation rates caused by a higher clastic input by rivers (Griffin, 1999). The Tortonian was characterized by a relatively arid eastern Mediterranean (Griffin, 1999).

Material and Methods

Material

The upper Miocene marine succession of the Pissouri Basin on Cyprus consists of an alternation of indurated carbonate rich and less indurated marlier beds and shows a distinct cyclicity (Krijgsman *et al.*, 2002; Kouwenhoven *et al.* 2006). The softer marly beds of the Pissouri section can be correlated to contemporaneous sapropelic layers throughout the Mediterranean (Kouwenhoven *et al.*, 2006).

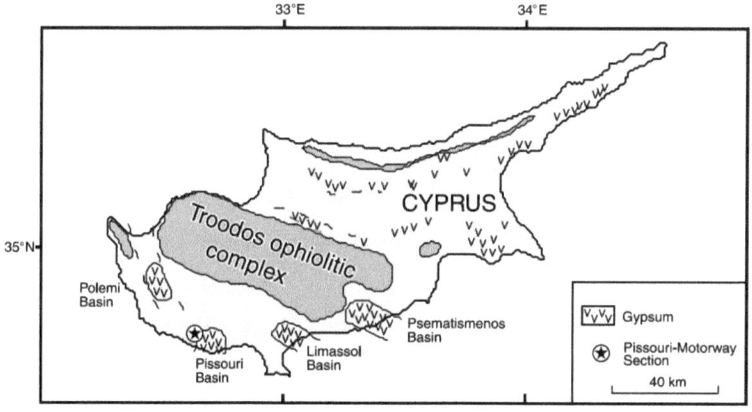

Fig. 1. Location of the Pissouri section on Cyprus.

Our record from Cyprus comprises the complete marine succession of the Pissouri Basin before, during and after the Messinian Salinity Crisis. The calcareous dinoflagellate data of this paper are derived from two land sections in the Pissouri Basin (Fig. 1), the Pissouri Motorway section (40 samples) (Krijgsmann *et al.*, 2002) and the Pissouri Village section (1 sample) (Orszag-Sperber *et al.* 1989, 2000; Rouchy *et al.*, 2001). The Pissouri Motorway section is located along the Limassol-Paphos motorway (Krijgsmann *et al.*, 2002) about 100 m west of the Pissouri village section, which is exposed at the eastern entry of

Pissouri (Rouchy *et al.*, 2001). Our samples comprise the interval between the late Tortonian at ~7.6 Ma (cycle VII) up to the base of the massive lower gypsum deposits at 5.98 Ma (cycle 1) and the first centimetres of the basal Pliocene, belonging to the Miocene-Pliocene 1 Biozone of Cita (1975) (MPL1), deposited under deep marine conditions (Rouchy *et al.*, 2001) (Fig. 2). The upper Messinian evaporitic and post-evaporitic interval (5.96-5.32 Ma) was not considered due to its essentially nonmarine character (e.g. Castradori, 1998; Rouchy *et al.*, 2001).

Except for the uppermost two cycles of the Tortonian, the late Tortonian and early Messinian cycles are thinner (around 2.6 cm/ky) than those of the late Messinian (around 5 cm/ky) (Kouwenhoven *et al.*, 2006) (Fig. 2). Consequently, from cycle 18 upwards the sedimentation rates are on average twice as high than in the cycles below. A detailed description of the Pissouri Motorway Section is given in Krijgsmann *et al.* (2002) and Kouwenhoven *et al.* (2006).

Methods

Samples consisting of 0.5 g dried sediment were disaggregated in 100 ml 0.6 % Soda solution. For samples with elevated organic matter contents (e.g. from sapropels), disaggregating and organic matter oxidation were combined in a 10 – 15 % H_2O_2 solution. To speed up the oxidation process samples were briefly heated to 50°C. Samples failing to disaggregate with these procedures were treated with repeated freezing and thawing in saturated sodium sulphate solution. After disaggregating, the suspensions were treated with ultrasound for less than 1 min to further separate the particles. The 20-75 μm fraction was isolated from the sample by washing with tap water over 75-μm and 20-μm sieves. Randomly selected pilot samples of the <20 μm and >75 μm appeared to contain no calcareous dinoflagellate cysts. To avoid carbonate dissolution, care was taken to perform all procedures in a slightly alkaline environment by adding a few drops of ammonia if necessary.

The 20-75 μm was washed into a scaled 20-ml test tube with lid. To reduce the surface tension and avoid contamination by fungi a few drops of ammonia and ethanol were added (Vink *et al.*, 2000). After a few hours settling, the water was removed with a pipette to a sample volume of 12 or 15 ml, depending on the amount of material. For quantitative scanning electron microscopic (SEM) analyses, a round cover glass (13 mm Ø) was applied to the tab using a double sided adhesive lead-tab. Subsequently, a 50 μl sub sample was applied to the SEM stub. This sub sample was taken at a depth of 1 cm below the surface of the thoroughly homogenized 12 ml or 15 ml sample using an Eppendorf micropipette. To get a better dispersion of the particles on the cover glass, surface tension was reduced by adding a small drop of ethanol. The solvents were evaporated by heating the stub to nearly 100°C. The stubs were sputtered with gold prior to SEM examination. SEM conditions were 15 – 20 kV, working distance 12 - 15 mm. Cysts were counted under the SEM, if necessary, multiple SEM stubs were analysed to reach a statistical relevant outcome. However, cyst concentrations were extremely low for the upper Messinian samples (see Appendix 2). The number of cysts per milligram of dry sediment counted on the SEM stubs was calculated according as:

$$Cysts/mg = C * S_t/(S_q * N * M),$$

where C is the number of counted cysts, S_t represents the total sample volume in ml (10 - 15), S_q represents the sub sample volume applied to the stub in ml (0.05 or 0.10), N is the number of counted stubs and M represents the dry weight of the examined sediment. Diversity has been calculated as follows:

$$H = - \sum_{i=1}^{n} n\,(p_i \ln p_i)$$

where H is the Shannon-Weaver Diversity Index, which increases with increasing heterogeneity of the sample. P_i is the relative abundance of individuals in the ith species, n is the total number of species in the community (richness), and ln is the natural logarithm.

Additionally, selected samples were analysed by polarised light microscopy with a gypsum plate (Janofske, 1996) to study the crystallographic orientation of the wall crystals. Especially for species with strong morphological similarities, e.g. *C. levantinum* and *P. tuberosa*, or morphological variations due to diagenetic overprint such as for *L. granifera*, the light optical investigations were used for species differentiation. Within a precession cycle the carbonate beds and marly intervals represent opposite environmental conditions. To enable analysis of the longer term environmental changes and to circumvent problems with the often unproductive carbonate beds and assure sample intercomparability, samples were analysed from the marly beds only. For taxonomic information on the examined species see Appendix 1.

Repository

The studied material is deposited in the collection of the Division of Historical Geology and Palaeontology, Department of Geosciences, University of Bremen, Germany.

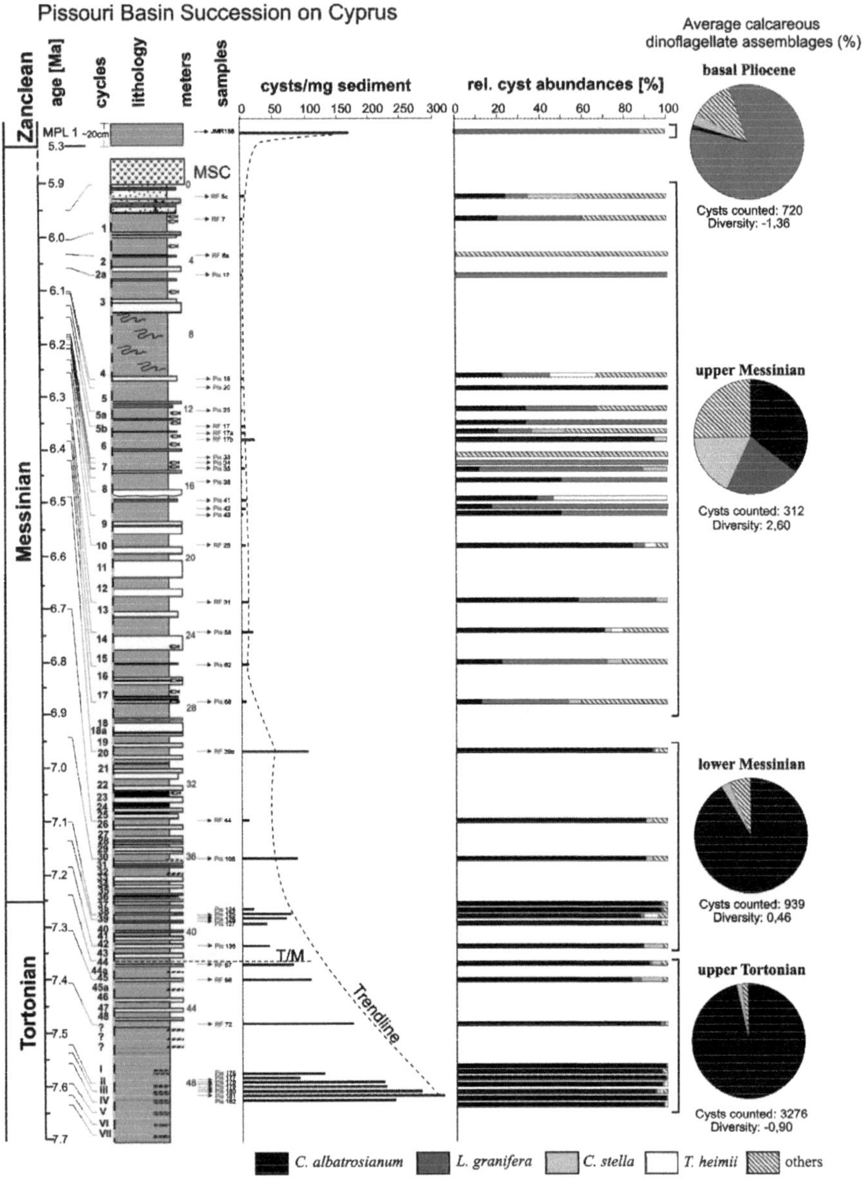

Fig. 2. Lithology (after Krijgsman *et al.*, 2002), age and cyst abundances (cysts mg_1 sediment) of the analysed samples and relative abundances of *C. albatrosianum*, *L. granifera*, *C. stella*, *T. heimii* and other cysts in the analysed samples against time. Lithology according to Krijgsman *et al.*, (2002). White levels correspond to more indurated beds, black levels to organic-rich layers (sapropels) and grey levels to softer blue-greyish and laminated marls; dotted levels represent the transitional interval to the evaporites composed of stromatolithic limestones; the 'v'-signed levels symbolize gypsum. Additionally, average calcareous dinoflagellate assemblages in relative abundances and average diversity during the upper Tortonian, lower Messinian, upper Messinian and the basal Pliocene are shown.

Results

Within the investigated interval 13, generally well preserved calcareous dinoflagellate species have been identified with strongly varying relative and absolute abundances through time (Fig. 2; App. 2). Cyst concentrations (cysts mg^{-1} of dry sediment) strongly vary (Fig. 2). Highest concentrations were recorded in the upper Tortonian samples (79 - 317cysts mg^{-1}), while lowest cyst concentrations occurred during the upper Messinian period (<21cysts mg^{-1}).

The diversity of our recorded dinoflagellate assemblages fluctuates between very low diversities during the upper Tortonian interval (on average: H= -0.90) and moderate to low diversities during the lower Messinian (on average: H= 0.46) (Fig. 2; App. 2). The dominating species of both time intervals is *C. albatrosianum* with relative abundances >83%, reaching maximum values of 100 % in the upper Tortonian samples (Fig. 2). On average *C. albatrosianum* distinctly dominates the upper Tortonian and lower Messinian assemblages with 96% and 92% respectively (Fig. 2).

The upper Messinian is marked by strongly varying relative cyst abundances and very low cyst concentrations (0.2 – 21 cysts mg^{-1}). As a result of this high variability, the average assemblage of this interval is much more diverse (H= 2.60) and, depending on the sample, dominated by *C. albatrosianum*, *L. granifera* or *Caracomia stella* (Fig. 2; App. 2). Compared to the upper Tortonian/lower Messinian, *C. albatrosianum* loses its distinct supremacy and becomes rarer with an average relative abundance of 34% during this interval (Fig. 2). *L. granifera* and *C. stella* clearly increase and temporally dominate the associations but with opposite trends (Fig. 2). *T. heimii*, *Pirumella parva*, *Calciodinellum operosum* and *Melodomuncula berlinensis* become more important (Fig. 2).

During the basal Pliocene the dominating species of the Tortonian/Messinian period are replaced by *L. granifera* with an 87% relative abundance (Fig. 2). This assemblage is marked by a relatively high cyst concentration (172 cysts mg^{-1}). The diversity is relatively low again (H= -1.36), even though the number of species (8) is comparatively high (Fig. 2). Species with slightly increased relative abundances are *Rhabdothorax* spp. (5 %) and *C. stella* (3 %). In addition *C. elongatum* occurs for the first time (2 % of the assemblage) (Fig. 2).

Light microscopic observations of the sibling species *C. levantinum* and *P. tuberosa* showed that only latter one was present.

Discussion

On the basis of calcareous dinoflagellates, the lithology and stratigraphy, four intervals have been distinguished, reflecting environmental conditions caused by the Messinian Salinity Crisis: 1.) upper Tortonian interval, 2.) lower Messinian interval, 3.) upper Messinian interval and 4.) basal Pliocene (Fig. 2).

The upper Tortonian is more or less monospecific and strongly dominated by *C. albatrosianum*, similar to the open tropical to subtropical Atlantic Ocean today (Vink, 2004). *C. albatrosianum* has been suggested to be a typical thermocline-dwelling species with a clear connection to warm, oligotrophic waters (Vink, 2004; Janofske & Karwath, 2000; Wendler *et al.*, 2002a, b). It also occurs in the eutrophic upwelling regions of the Arabian Sea (Wendler *et al.*, 2002a, b) and the equatorial and Benguela upwelling areas (Vink, 2004) but always with lower abundances than in the open ocean (Vink *et al.* 2003; Vink, 2004; Wendler *et al.*, 2002a, b).

One of the most striking features of the upper Tortonian interval is the extremely low diversity caused by the distinct dominance of *C. albatrosianum* (Fig. 2). Calcareous dinoflagellates which tolerate a wide range of environmental conditions have their highest diversity in coastal and more neritic regions (Karwath, 2000; Vink *et al.*, 2003; Vink, 2004). Low diversities and dominance of a single species such as during the Tortonian indicates relative stable oceanic conditions (Hildebrand-Habel & Willems, 1997; Kohring, 1997; Meier & Willems, 2003; Kohring *et al.*, 2005). This in contrast to planktic foraminifera and coccolithophora, where low diversities are typical characteristics of unstable, restricted environmental conditions (Kouwenhoven, 2000; Wade & Bown, 2005; Kouwenhoven *et al.*, 2006).

The lower Messinian interval has much less cysts mg^{-1} of dry sediment than the upper Tortonian one (Fig. 2). The assemblages are almost similar with only a slight decrease of *C. albatrosianum* and conversely a slightly increased diversity within the lower Messinian samples. These changes are mainly caused by increased relative abundances of *L. granifera* and *C. stella*. In the case of *L. granifera* this indicates a progressively riverine influence (Vink *et al.*, 2000; Wendler *et al.*, 2002b, c; Vink 2004) caused by the progressive separation of the Mediterranean Basin from the Atlantic Ocean. *C. stella* is as yet only known from warmer environments from low and middle latitudes of Miocene/Pliocene age (e.g. Fütterer, 1977a; Kohring, 1993a, b, 1997; Keupp & Kohring, 1999; Streng *et al.* 2002; Hildebrand-Habel & Streng, 2003) and from surface samples of the South Atlantic Ocean (Streng *et al.*, 2002).

Higher relative abundances of *C. stella* are associated with more shelfward environments (Keupp & Kohring, 1999; this work) but with a distribution pattern opposite to *L. granifera* (Fig. 2). Furthermore, higher concentrations of *C. stella* in the Mediterranean upper Miocene / Pliocene (Keupp & Kohring, 1993, 1999; Kohring, 1993b, 1997; this work) occur together with *C. albatrosianum*, representatives of the *edgarii*-group and *P. tuberosa*, implying similar environmental preferences (Tab. 1, Fig. 2). Hence, higher abundances of *C. stella* are probably indicative for oligotrophic, coastal warm waters with normal or slightly increased salinities.

Thus, the modifications of the lower Messinian interval possibly already reflect the beginning of more instable and restricted conditions caused by the Messinian salinity crisis. Importantly, this first notable shift in calcareous dinoflagellate assemblages already occurs shortly before the Tortonian-Messinian boundary at 7.3 Ma (Fig. 2; sample RF 67) and predates the disappearance of a group of open marine, deep-water benthic foraminifera taxa at 7.167 Ma in the Pissouri basin (Kouwenhoven *et al.*, 2006).

The third, upper Messinian, interval is much more diverse (H = 0.58-4.61) but possesses only few cysts. The dinoflagellate assemblages reveal striking variations in species abundances (Fig. 2) which are interpreted as to reflect fluctuating salinities, nutrients and temperatures towards the evaporitic stage. All samples show an inverse relationship between *L. granifera* on one hand and *C. stella* and to an even larger extent *C. albatrosianum* on the other. Abundance peaks of *L. granifera*, together with a decrease of *C. albatrosianum* and *C. stella* are likely the response to enhanced continental runoff, resulting in nutrient enriched surface waters with reduced salinities. High abundances of *C. albatrosianum* can be interpreted as a temporary re-establishment of normal marine oligotrophic conditions. In contrast to this the sporadic occurrence of the more neritic *Bicarinellum tricarinelloides*, *Melodomuncula berlinensis*, *Pirumella parva* and *Rhabdothorax* spp. indicate temporary shallowing of the basin. On average the upper Messinian assemblage is much more diverse (H = 2.6) and reflects very unstable, more neritic conditions with varying fluvial input. Additionally, frequent occurrence of ascidian sclerites emphasise the more neritic character of this interval (Fütterer, 1977).

The basal Pliocene assemblage reflects the restoration of normal marine conditions just after the replenishment of the Mediterranean Basin (Spezzaferri *et al.*, 1998; Rouchy *et al.*, 2001). However, a strong change in the association, compared to the Miocene assemblages, took place: *L. granifera* clearly dominates the assemblage and obviously replaced *C. albatrosianum*. It is hypothesized that a low salinity, high nutrient layer of freshwater must have spread out over the dense, saline, waters. Rouchy *et al.* (2001) and Iaccarino *et al.* (1999) also observed a continued and significant freshwater input in the Mediterranean Basin, which probably affected particularly proximal land areas such as the Pissouri Basin in the eastern Mediterranean (Iaccarino *et al.*, 1999; Rouchy *et al.*, 2001). These observations support the hypothesis that simultaneous to the salinity crisis, the Mediterranean climate changed from a more arid Tortonian to a humid upper Messinian and basal Pliocene (Diester-Haass *et al.*, 1998; Fauquette *et al.* 1998), marked by high monsoonal activity (Griffin, 1999). Such intensified monsoonal precipitation also occurs during orbital precession minima which are in turn associated to strong Northern Hemisphere (NH) summer insolation, resulting in increased discharge from the Nile (e.g. Rossignol-Strick, 1985; Hilgen 1991; Diester-Haass *et al.*, 1998). Although Mediterranean sapropels have been formed during precession minima (Wehausen & Brumsack, 1998; Lange *et al.*, 1999; Meier *et al.*, 2004) precession minima at the basal Pliocene (Rossignol-Strick, 1985; Hilgen, 1991; Castradori, 1998; Steenbrink *et al.*, 2006; Van der Laan *et al.*, 2006) did not lead to sapropel formation in the Pissouri Basin.

Comparison to other Neogene Mediterranean assemblages

The species spectrum of calcareous dinoflagellates from the Pissouri section agrees well with that reported by most other Neogene studies from the Mediterranean so far (Keupp & Kohring, 1993, 1999; Kohring, 1993, 1997; Tab. 1). Only some rare species such as *Cervisiella saxea* and *Calcipterellum colomi*, could not be found in the Pissouri samples. One exception is *Orthopithonella sicelis*, which is only known from a poorly preserved lower Pliocene section on Sicily, where it dominates (26%) the calcareous dinoflagellate assemblage (Kohring, 1993; Tab. 1, included in the *edgarii*-group). In our opinion this basal Pliocene assemblage from Sicily reflects a more neritic environment with *O. sicelis* as the dominating species and abundant *P. parva*. In our material *O. sicelis* could not be found whereas *P. parva* is scarce (Fig. 2). An obvious explanation is, that the basal Pliocene of the Pissouri Basin represents a deeper marine environment (Rouchy *et al.*, 2001) not suitable for these taxa.

However, a comparison of the Pliocene dinoflagellate assemblage from Sicily (Kohring, 1993) and Messinian assemblages from Algeria (Keupp & Kohring, 1993) with our upper Messinian assemblage shows more agreement in relative abundances and the species spectrum (Tab. 1). All three assemblages are relatively high diverse and share most major members (*L. granifera*, *C. albatrosianum*, *C. stella*), however there are deviations in relative abundances and lower contributions of *C. albatrosianum* and *C. stella* in the Algerian and Sicilian assemblages (Tab.1). The striking difference between the upper Miocene assemblage of the Pissouri section and those of the Algerian and Sicilian ones is the high abundance of representatives of the *edgarii*-group in the latter sections (Tab. 1), which gives them a stronger neritic character. Furthermore, the Sicilian assemblage is characterized by a slightly higher amount (10%) of *O. tuberosa* (now *Pirumella tuberosa)*, reflecting elevated salinity and more oligotrophic conditions (Vink, 2004). Therefore, the dilution and eutrophication via fluviatile input might have been less significant.

A relatively shallow warm water environment is proposed with mesotrophic and mesohaline conditions for the upper Miocene Algerian and basal Pliocene Sicilian sedimentation areas. Although Kohring (1993) and Keupp & Kohring (1993) also assume warm water conditions and reveal the near shore character of the *edgarii*-group a concluding interpretation of the environment is missing. Abundance variations of *C. albatrosianum* and *L. granifera* (formerly *Sphaerodinella albatrosiana* and *Orthopithonella granifera*, respectively) are merely seen as fluctuating Atlantic influences. We interpret these variations as a fluctuating terrigenous runoff and associated salinity and nutrient variations.

Regardless of the different stratigraphical and palaeogeographical positions our basal Pliocene assemblage and the upper Miocene (Serravallian) assemblage of Kohring (1997) from Malta (Blue Clay Formation) agree well (Tab. 1). Both assemblages are more or less monospecific with a clear dominance of *L. granifera* (87% and 80% respectively) reflecting a strong fluviatile influence.

Keupp & Kohring (1999) describe a calcareous dinoflagellate assemblage from the upper Miocene of Crete (Episkopi) (Tab. 1) which differs from the aforementioned associations by the absence of *L. granifera* and elevated amounts of *C. albatrosianum*. Again, due to its interpretation as a typical pelagic open marine species (Kohring, 1993, 1997) Keupp & Kohring (1999) interpret the absence of *L. granifera* as a decreasing Atlantic influence. We further attribute the absence of *L. granifera* to missing fluviatile influence.

Our results agree well with other microfossil studies of the Pissouri Basin (Rouchy *et al.* 2001; Krijgsmann *et al.*, 2002; Kouwenhoven *et al.*, 2006). Only the calcareous dinoflagellates reported by Kouwenhoven *et al.*, (2006) differ significantly from our data; this relates to the use of different methodologies and taxonomic concepts.

Our interpretation of a warm, oligotrophic environment for the upper Tortonian interval correlates well with the distribution pattern of coccoliths and planktic foraminifera (Kouwenhoven *et al.*, 2006). These authors infer a transition from a cool-water, high-productivity environment (abundance maximum of *Coccolithus pelagicus*) to higher sea surface temperatures (SST) and oligotrophic conditions just below our lowermost sample at 7.5 Ma. This shift is also apparent from the decrease of the cold-water indicating planktic foraminifera (Neogloboquadrinids and *Globorotalia* spp.) and an increase in the subtropical and oligotrophic *Globigerinoides* spp. (Kouwenhoven *et al.*, 2006).

Pissouri Basin associations versus recent Mediterranean assemblages

The calcareous dinoflagellate assemblages from the Pissouri Basin are very unlike those of the recent Mediterranean Sea. Although most of the main upper Miocene to lower Pliocene calcareous dinoflagellate species are extant, some species of the modern Mediterranean are missing in our record and vice versa. The most relevant species in this context are *C. levantinum* and *P. parva*. *C. levantinum* today dominates the western Mediterranean associations but distinctly decreases eastwards (Meier & Willems, 2003). Currently it is not known when *C. levantinum* started its extension into the eastern Mediterranean Basin. *P. parva*, a representative of the *edgarii*-group is lacking in the modern Mediterranean. This species disappeared possibly somewhere during the late Pliocene, possibly in relation to the onset of the northern Hemisphere glaciations (Streng *et al.*, 2004). *P. parva* has been suggested to prefer near shore environments and warm waters (Hildebrand-Habel & Willems, 2000; Hildebrand-Habel & Streng, 2003).

The dominance of *C. albatrosianum* in the late Tortonian to early Messinian assemblages has no equivalent in the modern Mediterranean. Today this species accounts for less than 5% of the Mediterranean associations (Meier & Willems, 2003), similar to its abundance in our Pliocene record. It is proposed that conditions similar to those in modern oceans with high amounts of *C. albatrosianum* (Wendler *et al.*, 2002a; Tanimura & Shimada, 2004; Vink, 2004), prevailed in the Mediterranean during the upper Tortonian and lower Messinian.

C. elongatum, a common species in the eastern Mediterranean today, occurs for the first time in our basal Pliocene sample, might be seen as an Atlantic newcomer. *T. heimii* is the most abundant calcareous dinoflagellate species in the present day Mediterranean Sea (e.g. Zonneveld *et al.*, 2001; Meier & Willems, 2003). This species, which can be traced back to the K/T boundary (Hildebrandt-Habel *et al.* 1999), occurs in our material only sporadically and with very low relative and absolute abundances (<2 cysts mg^{-1}). The only exceptions are in two of the upper Messinian samples (Fig. 2) where this species accounts for 22 and 54% of the assemblage. However, in total this species remains rare over the Messinian Salinity Crisis and after the restoration of marine conditions with the beginning of the Pliocene. Until now it remains unclear why, and when *T. heimii* started to dominate the Mediterranean calcareous dinoflagellate assemblages.

L. granifera and *C. stella* which increase in progressivity to the Messinian Salinity Crisis, are either very scarce or not present (*C. stella*) in the modern Mediterranean (Meier & Willems, 2003).

Only one specimen of *L. granifera* was reported from a survey of surface samples taken throughout the Mediterranean by Meier & Willems (2003). However, this species is abundant in the S1 sapropel (~8-30% of the assemblages) (Meier *et al.*, 2004) an environment which is therefore closer to our basal Pliocene environment. The other species linked to the S1 sapropel, *C. levantinum*, is not present in our samples. Compared to the present Mediterranean, which is characterized by oligotrophic, high saline water masses with strong east-west gradients (Meier *et al.*, 2004), the Mediterranean during S1 deposition was marked by enhanced freshwater discharge and consequently increased nutrient concentration (Meier *et al.*, 2004). A similar situation, but far much severe, is proposed to have occurred during the basal Pliocene in the eastern Mediterranean when huge amounts of freshwater spread out over the re-establishing Mediterranean (e.g. Diester-Haass *et al.*, 1998; Rouchy *et al.*, 2001). The species that apparently benefits most from this configuration in both settings is *L. granifera*. Interestingly, this did not lead to sapropel formation in the case of the lowermost Pliocene, suggesting sufficient ventilation of the deeper water masses.

The low relative abundance of *C. albatrosianum*, mentioned above, a feature of this basal Pliocene assemblage is in common with the modern Mediterranean ones. Furthermore, *L. urania* and *C. elongatum*, which are common in the eastern Mediterranean today, peak (*L. urania*) or appear for the first time (*C. elongatum*) in the Pliocene material.

L. urania and *C. elongatum* have a strong affinity to oligotrophic waters with elevated salinities, such as in the modern eastern Mediterranean, where these species form large parts of the calcareous dinoflagellate assemblages (Meier *et al.*, 2004). Today *L. urania* is

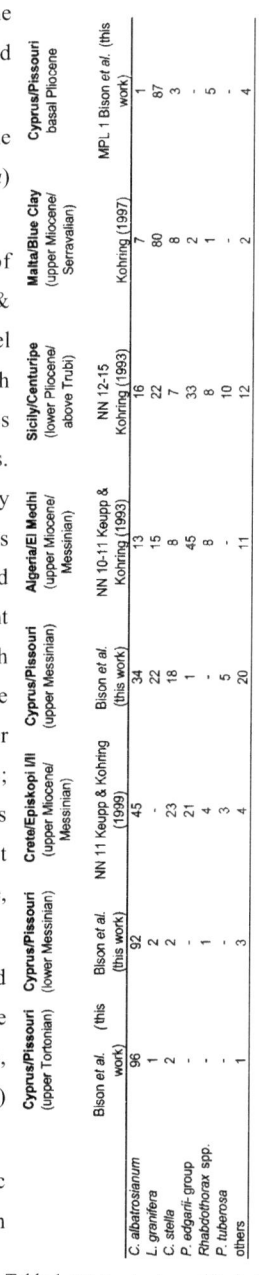

	Cyprus/Pissouri (upper Tortonian)	**Cyprus/Pissouri** (lower Messinian)	**Crete/Episkopi I/II** (upper Miocene/ Messinian)	**Cyprus/Pissouri** (upper Messinian)	**Algeria/El Medhi** (upper Miocene/ Messinian)	**Sicily/Centuripe** (lower Pliocene/ above Trubi)	**Malta/Blue Clay** (upper Miocene/ Serravalian)	**Cyprus/Pissouri** basal Pliocene
	Bison *et al.* (this work)	Bison *et al.* (this work)	NN 11 Keupp & Kohring (1999)	Bison *et al.* (this work)	NN 10-11 Keupp & Kohring (1993)	NN 12-15 Kohring (1993)	Kohring (1997)	MPL 1 Bison *et al.* (this work)
C. albatrosianum	96	92	45	34	13	16	7	1
L. granifera	1	2	-	22	15	22	80	87
C. stella	2	2	23	18	8	7	8	3
P. edgarii-group	-	-	21	1	45	33	2	-
Rhabdothorax spp.	-	1	4	-	8	8	1	5
P. tuberosa	-	-	3	5	-	10	-	-
others	1	3	4	20	11	12	2	4

Table 1. Relative abundances (%) of calcareous dinoflagellate species from selected Neogene Mediterranean localities discussed in the text.

almost entirely restricted to the high saline (~39psu) and oligotrophic eastern Mediterranean Sea and only few specimens have been recorded from the Tyrrhenian Sea (Meier *et al.*, 2002; Meier & Willems, 2003; Meier *et al.*, 2004). Even though *L. urania* and *C. elongatum* are present within our basal Pliocene samples they remain scarce. *L. urania* is restricted to the eastern Mediterranean today, it has also been reported from the Miocene Indian Ocean (Streng *et al.*, 2004), but not from the Atlantic. This gives the species a Tethyan rather than an Atlantic signature. We therefore suggest, in agreement with Meier & Willems (2003), that some specimens of *L. urania* survived the Salinity Crisis within the Mediterranean. It is unclear when *L. urania* established its dominant position in the eastern Mediterranean.

Conclusions

The impact of the Messinian Salinity Crisis on Mediterranean calcareous dinoflagellate assemblages was analysed from the Pissouri Basin on Cyprus. Time slices before, during and after the Salinity Crisis were investigated. The main objective of this paper was to investigate longer-term palaeoenvironmental changes in relation to the Messinian Salinity Crisis in the Eastern Mediterranean based on calcareous dinoflagellates. The dominance of *C. albatrosianum* and its high absolute and relative abundance during the upper Tortonian and lower Messinian is unlike modern Mediterranean Sea associations. It is more similar to the modern tropical Atlantic (Vink *et al.*, 2002b). Changing environmental conditions caused by the separation of the Mediterranean Sea from the Atlantic Ocean, superimposed by short term (precession controlled) variability, led to the replacement of C. *albatrosianum* dominated assemblages by *L. granifera* dominated ones after the Salinity Crisis. At the same time the species diversity changed from low diverse, almost monospecific associations during the upper Tortonian and lower Messinian through relatively high diverse associations during the upper Messinian to low diversity ones again during the basal Pliocene. With the re-filling of the Mediterranean Basin, just after the Miocene / Pliocene Boundary, the Atlantic element *C. elongatum* occurs first time and contributes to a more modern character of the Pliocene associations. The basal Pliocene calcareous dinoflagellate assemblage most closely resembles those of the Holocene S1 sapropel but still differs from the present Mediterranean ones. The final changes to current Mediterranean calcareous dinoflagellate associations must have taken place after the disappearance of the more or less monospecific earliest Pliocene association. Finally it can be said, that the Messinian Salinity Crisis did not lead to a permanent removal of oceanic calcareous dinoflagellate taxa from the Pissouri Basin. Most of the dinoflagellate cyst species of this period are still present in the modern Mediterranean and they can be interpreted as re-immigrants to the Mediterranean from the Atlantic. The exception is *L. urania*, which must have survived the salinity crisis in the Mediterranean.

Acknowledgements

The authors thank Jacob Nsiah (Bremen University) for preparing the samples and Hartmut Mai (Bremen University) for assistance and advice during the SEM analysis. Wout Krijgsman (Utrecht University) is thanked for making available the samples and Jean-Marie

Rouchy and Rachel Flecker for providing additional sample material. Special thanks go to Erik Snel, Erwin van der Laan and Tanja Kouwenhoven, (Utrecht University) for their constant assistance with the procurement of the samples. We also thank all members of the working group of Historical Geology and Palaeontology of Bremen University for their general assistance and openness to discussion. The authors are particularly grateful to the two reviewers Sebastian Meier and Helmut Keupp for valuable suggestions and criticisms that helped to improve the manuscript. The editor, John Gregory, is thanked for his critical review of the manuscript. Financial support by the German Research Foundation is gratefully acknowledged.

Plate 1

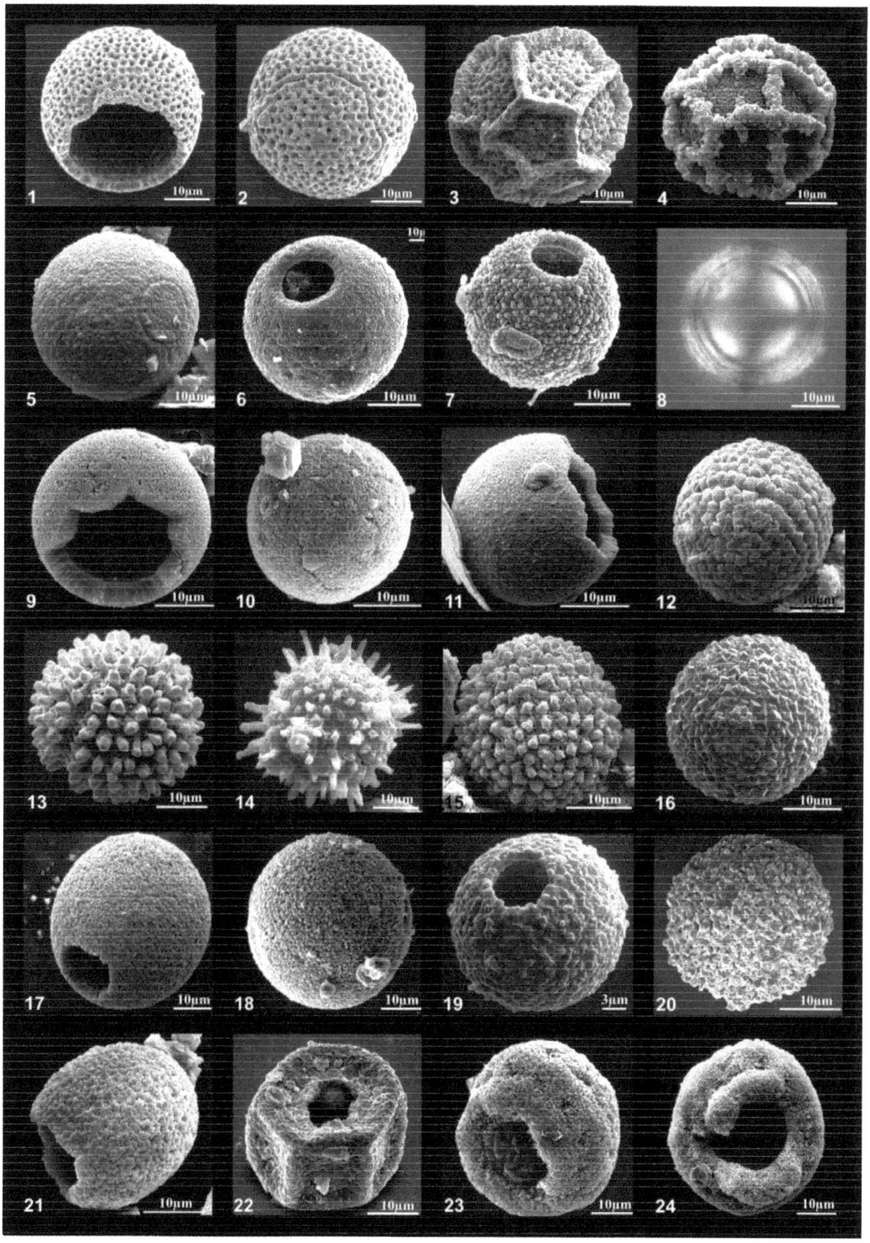
1
10µm
2
10µm
3
10µm
4
10µm
5
10µm
6
10µ
10µm
7
10µm
8
10µm
9
10µm
10
10µm
11
10µm
12
13
10µm
14
15
10µm
16
10µm
17
10µm
18
10µm
19
3µm
20
10µm
21
10µm
22
10µm
23
10µm
24
10µm

Explanation of Plate 1

Scanning electron (SEM) images and one light microscope image.

figs 1-2: *Calciodinellum albatrosianum*. **1** (sample PIS182), upper Tortonian, apical view, open cyst with polygonal archaeopyle. **2** (sample PIS182), upper Tortonian, apical view, closed cyst with polygonal delineated operculum.

figs 3-4: *Calciodinellum operosum*. **3** (sample RF17a), upper Messinian, paratabulated cyst with well developed crystal ridges and large pores. **4** (sample RF17a), upper Messinian, lateral view, closed cyst with blocky crystallite ridges and reduced pores.

figs 5-8: *Leonella granifera*. SEM pictures of *L. granifera*. **5** (sample JMR158), basal Pliocene, side apical view; spherical cyst with round operculum. **6**, (sample RF17a), upper Messinian, Side apical view, cyst with open archaeopyle. **7**, (sample JMR158), basal Pliocene, side view, cysts with open archaeopyle and coarser surface crystallites. **8**, (sample JMR158), Light optical view of *L. granifera* under crossed nicols.

figs 9-11: *Caracomia stella* Streng *et al*., 2002. **9**, (sample RF17a), upper Messinian, apical view, open cyst with large polygonal archaeopyle. **10**, (sample JMR158), basal Pliocene, side apical view of closed cyst, with delineated archaeopyle suture, **11**, (sample RF17a), upper Messinian, side view of open cyst.

fig. 12: *Lebessphaera urania* (sample PIS136), upper Tortonian. Lateral view of closed cyst, operculum suture visible in the upper part.

figs 13-15: *Rhabdothorax* spp. **13** (sample PIS126), upper Messinian, spherical closed cyst with shorter blocky spines. **14**, (sample PIS108), upper Messinian, closed cyst with long spines. **15** (sample PIS126), upper Messinian, spherical cyst with reduced blocky spines.

fig. 16: *Pernambugia tuberosa* (sample RF17a), upper Messinian. Closed spherical cyst with pyramid like cyst surface crystals.

figs 17-18: *Pirumella parva* (sample PIS68), upper Messinian. **17**, side apical view, elongated cyst with small round archaeopyle. **18**, sphaerical closed cyst, no operculum suture visible.

fig. 19: *Thoracosphaera heimii* (sample PIS41), upper Messinian. Spherical cyst with small round archaeopyle, irregularly surface crystals.

fig. 20: *Pirumella loeblichii* (sample PIS180), upper Tortonian. Sphaerical cyst with irregularly arranged surface crystals.

fig. 21: *Calciodinellum elongatum* (sample JMR158), basal Pliocene. Elongated cyst with open paratabulated archaeopyle.

fig. 22: *Melodomuncula berlinensis* (sample RF17a), upper Messinian. Lateral apical view of open cyst.

figs 23-24: *Bicarinellum tricarinelloides* (sample RF17a), upper Messinian. **23**, side apical view of open cyst. **24**, same cyst, apical view.

Appendix 1

Annotated listing of calcareous dinoflagellate taxa found in the investigated material

Division **Dinoflagellata** (Bütschli, 1885) Fensome *et al.*, 1993

Subdivision **Dinokaryota** Fensome *et al.*, 1993

Class **Dinophyceae** Pascher, 1914

Subclass **Peridiniphycidae** Fensome *et al.*, 1993

Order **Peridiniales** Haeckel, 1894

Suborder **Peridiniineae** Autonym

Family **Peridiniaceae** Ehrenberg, 1831

Subfamily **Calciodinelloideae** Fensome *et al.*, 1993

Bicarinellum tricarinelloides Versteegh, 1993 (pl. 1, figs 23-24)

Calciodinellum albatrosianum (Kamptner, 1963) Janofske & Karwath, 2000 (pl. 1, figs 1-2)

Calciodinellum elongatum (Hildebrand-Habel *et al.*, 1999) Meier *et al.*, 2002 (pl. 1, fig. 21)

Calciodinellum operosum Deflandre, 1947 (pl. 1, figs 2-4)

Caracomia stella (Gilbert & Clark, 1983) Streng *et al.*, 2002 (pl. 1, figs 9-11)

Lebessphaera urania Meier *et al.*, 2002 (pl. 1, fig. 12)

Melodomuncula berlinensis Versteegh, 1993 (pl. 1, fig. 22)

Pernambugia tuberosa (Janofske & Karwath, 2000) Hildebrand-Habel *et al.*, 1999 (pl. 1, fig. 16)

Pirumella parva (Bolli, 1974) Lentin & Williams, 1993 (pl. 1, fig 17-18)

Pirumella loeblichii (Bolli, 1974) Lentin & Williams, 1993 (pl. 1, fig. 20)

Rhabdothorax spp. (Kamptner, 1937) Kamptner, 1958 (pl. 1, figs 13-15)

Order **Thoracosphaerales** Tangen in Tangen *et al.*, 1982

Family **Thoracosphaeraceae** Schiller, 1930 emend. Tangen in Tangen *et al.*, 1982

Thoracosphaera heimii (Lohmann, 1920) Kamptner, 1944 (pl. 1, fig. 19)

Leonella granifera (Fütterer, 1977) Janofske & Karwath, 2000 (pl. 1, figs 5-8)

According to the new phylogenetic idears of Gottschling *et al.* (2005) and the close relationship of *Leonella* and *Thoracpsphaera*, *L. granifera* is grouped under Thoracospherales rather than under Calciodinelloids as previously. The light optical appearance of our cysts is identical to that of typical *L. granifera*. However, upon SEM analysis it appears that most of our *L. granifera* cysts differ from the type in the absence of pores in the outer cyst surface and by a characteristic wall surface with epitaxially grown crystals. Since the whole spectrum of intermediate forms is also observed, we iterpreted the observed morphological variation as a slight diagenetic overprint. This morphological variation of *L. granifera* was also observed in other Neogene Mediterranean samples (e.g. Kohring, 1993; Keupp & Kohring, 1993).

Appendix 2

Count data and calculated diversity data for all analysed samples.

Appendix 2

	C. albatrosianum		*C. operosum*		*C. elongatum*		*C. stella*		*L. granifera*		*L. urania*		*M. berlinensis*		*B. tricarinelloides*		*P. parva*		*P. tuberosa*		*Rhabdothorax* spp.		*P. loeblichii*		*T. heimii*		not identified		total no of	
sample	No	%	No	%	No	%	No	%	No	%	No	%	No	%	No	%	No	%	No	%	No	%	No	%	No	%	No	%	Dinos	Diversity
JMR 158	7	1	1	0	11	2	18	2	626	87	11	2	2	0	0	0	0	0	0	0	38	5	0	0	0	0	6	1	721	-1.37
RF 5c	7	24	1	3	0	0	7	24	3	10	2	7	0	0	0	0	0	0	0	0	0	0	0	0	0	0	9	31	29	2.82
RF 7	1	20	0	0	0	0	0	0	2	40	0	0	1	20	0	0	0	0	0	0	0	0	0	0	0	0	1	20	5	4.33
RF 8a	0	0	0	0	0	0	0	0	0	0	0	0	0	0	1	100	0	0	0	0	0	0	0	0	0	0	0	0	1	4.61
PIS 12	0	0	0	0	0	0	2	100	0	0	0	0	0	0	0	0	0	0	0	0	0	0	0	0	0	0	0	0	2	3.91
PIS 18	2	22	0	0	0	0	2	22	0	0	0	0	0	0	0	0	0	0	0	0	0	0	0	0	2	22	3	33	9	3.78
PIS 20	2	100	0	0	0	0	0	0	0	0	0	0	0	0	0	0	0	0	0	0	0	0	0	0	0	0	0	0	2	3.91
PIS 25	1	33	1	33	0	0	1	33	0	0	0	0	0	0	0	0	0	0	0	0	0	0	0	0	0	0	0	0	3	4.61
RF 17	1	33	0	0	0	0	2	67	0	0	0	0	0	0	0	0	0	0	0	0	0	0	0	0	0	0	0	0	3	4.14
RF 17a	5	20	0	0	0	0	4	16	4	16	0	0	2	8	8	32	0	0	1	4	1	4	0	0	0	0	0	0	25	3.12
RF 17b	66	94	0	0	0	0	4	6	0	0	0	0	0	0	0	0	0	0	0	0	0	0	0	0	0	0	0	0	70	0.58
PIS33	0	0	0	0	0	0	0	0	0	0	0	0	0	0	0	0	0	0	1	100	0	0	0	0	0	0	0	0	1	4.61
PIS34	0	0	0	0	0	0	0	0	1	100	0	0	0	0	0	0	0	0	0	0	0	0	0	0	0	0	0	0	1	4.61
PIS35	1	11	0	0	0	0	1	11	7	78	0	0	0	0	0	0	0	0	0	0	0	0	0	0	0	0	0	0	9	3.09
PIS38	1	50	0	0	0	0	0	0	1	50	0	0	0	0	0	0	0	0	0	0	0	0	0	0	0	0	0	0	2	4.61
PIS41	5	38	0	0	0	0	1	8	0	0	0	0	0	0	0	0	0	0	0	0	0	0	0	0	7	54	0	0	13	2.94
PIS42	1	17	0	0	0	0	5	83	0	0	0	0	0	0	0	0	0	0	0	0	0	0	0	0	0	0	0	0	6	3.26
PIS43	1	50	0	0	0	0	0	0	1	50	0	0	0	0	0	0	0	0	0	0	0	0	0	0	0	0	0	0	2	4.61
RF 25	15	83	0	0	0	0	0	0	1	6	1	6	0	0	0	0	0	0	0	0	0	0	0	0	1	6	0	0	18	2.35
RF 31	11	58	0	0	0	0	1	5	7	37	0	0	0	0	0	0	0	0	0	0	0	0	0	0	0	0	0	0	19	2.5
PIS 58	23	70	0	0	0	0	1	3	0	0	0	0	0	0	0	0	1	3	0	0	1	3	0	0	2	6	5	15	33	2.13
PIS 62	9	21	1	2	0	0	3	7	21	50	0	0	3	7	0	0	2	5	0	0	1	2	0	0	0	0	2	5	42	2.39
PIS 68	2	12	0	0	0	0	1	6	7	41	0	0	0	0	0	0	3	18	0	0	0	0	0	0	0	0	4	24	17	3.2
RF 39a	212	97	0	0	0	0	1	0	4	2	0	0	0	0	1	0	0	0	0	0	0	0	0	0	1	0	0	0	219	-0.61
RF 44	19	83	1	4	0	0	1	4	2	9	0	0	0	0	0	0	0	0	0	0	0	0	0	0	0	0	0	0	23	2.11
PIS 108	172	93	0	0	0	0	0	0	2	1	2	1	0	0	0	0	0	0	0	0	2	1	0	0	3	2	4	2	185	-0.25
PIS 124	27	90	0	0	0	0	1	3	0	0	1	3	0	0	0	0	1	3	0	0	0	0	0	0	0	0	0	0	30	1.64
PIS 125	123	97	0	0	0	0	0	0	1	1	2	2	0	0	0	0	0	0	0	0	1	1	0	0	0	0	0	0	127	-0.07
PIS 126	102	87	1	1	0	0	0	0	2	2	0	0	0	0	0	0	0	0	0	0	3	3	0	0	8	7	1	1	117	0.39
PIS 127	62	97	0	0	0	0	0	0	0	0	0	0	0	0	0	0	0	0	0	0	0	0	1	2	1	2	0	0	64	0.61
PIS136	154	89	0	0	0	0	16	9	0	0	4	2	0	0	0	0	0	0	0	0	0	0	0	0	0	0	0	0	174	-0.14
RF67	151	92	1	1	0	0	7	4	2	1	3	2	0	0	0	0	1	1	0	0	0	0	0	0	0	0	0	0	165	-0.10
RF68	188	83	0	0	0	0	23	10	10	4	3	1	0	0	0	0	0	0	0	0	0	0	0	0	0	0	2	1	226	-0.19
RF 72	352	96	0	0	0	0	2	1	7	2	0	0	0	0	0	0	1	0	0	0	0	0	0	0	0	0	3	1	365	-1.10
PIS 176	212	99	0	0	0	0	1	0	0	0	0	0	0	0	0	0	0	0	0	0	1	0	0	0	0	0	0	0	214	-0.70
PIS 177	147	97	0	0	0	0	0	0	0	0	0	0	0	0	0	0	3	2	0	0	0	0	0	0	0	0	1	1	151	-0.27
PIS 178	372	99	0	0	0	0	0	0	0	0	2	1	0	0	0	0	1	0	0	0	0	0	0	0	0	0	0	0	375	-1.27
PIS 179	370	97	0	0	0	0	2	1	0	0	3	1	0	0	0	0	1	0	2	1	0	0	0	0	2	1	0	0	380	-1.17
PIS 180	444	94	0	0	0	0	8	2	1	0	0	0	0	0	0	0	3	1	0	0	0	0	8	2	6	1	0	0	470	-1.25
PIS 181	527	100	1	0	0	0	0	0	0	0	0	0	0	0	0	0	0	0	0	0	0	0	0	0	0	0	0	0	528	-1.65
PIS 182	395	98	0	0	0	0	1	0	0	0	1	0	0	0	0	0	1	0	0	0	0	0	4	1	0	0	0	0	402	-1.28
																												sum:	5248	

References

Bianchi, C. N., Morri, C., 2000. Marine biodiversity of the Mediterranean Sea : situation, problems and prospects for the future research. *Marine Pollution Bulletin*, **40**(5): 367-376.

Blanc, P.-L., 2002. The opening of the Plio-Quaternary Gibraltar Strait: assessing the size of a cataclysm. *Geodinamica Acta*, **15**: 303–317.

Bouchet, P., Taviani, M., 1992. The Mediterranean deep-sea fauna: pseudopopulations of Atlantic species? *Deep Sea Research*, **39**: 169-184.

Bouillon, J., Medel, M.D., Pagos, F., Gili, J.-M., Boero, F., Gravili, C., 2004. Fauna of the Mediterranean Hydrozoa. *Scientia Marina* (Barcelona), *Consejo Superior de Investigaciones Cientificas*, Institut de Ciènces del Marina, Barcelona, Spain, **68**(2): 5-449.

Butler, R.W.H., McClelland, E., Jones, R.E., 1999. Calibrating the duration and timing of the Messinian salinity crisis in the Mediterranean: linked tectonoclimatic signals in thrust-top basins in Sicily. *Journal of the Geological Society, London,* **156**: 827-835.

Carnevale, G., Landini, W., Sarti, G. 2006. Mare versus Lago-mare: marine fishes and the Mediterranean environment at the end of the Messinian Salinity Crisis. *Journal of the Geological Society, London,* **163**: 75-80.

Castradori, D., 1998. Calcareous nannofossil in the basal Zanclean of the eastern Mediterranean Sea: Remarks on paleoceanography and sapropel formation. *Proceedings of the Ocean Drilling Program, Scientific Results,* **160**: 113-123.

Cita, M.B., 1975. Studi sul Pliocene e gli strati di passagio del Miocene al Pliocene, VIII. Planktonic foraminiferal biozonation of the Mediterranean Pliocene deep sea record: a revision. *Rivista Italia di Paleontologia e Stratigrafia*, **81**: 527-544.

Dale, B., 1992. Dinoflagellate contribution to the open ocean sediment flux. *In*: *Honjo*, S. (Ed.). Dinoflagellate contribution to the Deep Sea. *Ocean Biocoenosis Series*, Oceanographic Institution, Woods Hole, Massachusetts, **5**: 1-32.

De Lange, G. J., van Santvoort, P. J. M., Langereis C., Thomson, J., Corselli, C., Michard, A. Rossignol-Strick, M., Paterne, M., Anastasakis, G., 1999. Palaeo-environmental variations in eastern Mediterranean sediments: A multidisciplinary approach in a prehistoric setting, *Progress in Oceanography*, **44**: 369–386.

Diester-Haass, L., Robert, C., Chamley, H., 1998. Paleoproductivity and climate variations during sapropel deposition in the eastern Mediterranean Sea. *Proceedings of the Ocean Drilling Program, Scientific Results,* **160**: 227-248.

Domingues, V.S, Bucciarelli, G., Almada, V.C., Bernardi, G., 2005. Historical colonization and demography of the Mediterranean Damselfish, *Chromis chromis*. *Molecular Ecology*, **14**(13): 4051-4063.

Emig, C., Geistdoerfer, P. 2004. The Mediterranean deep-sea fauna: historical evolution, bathymetric variations and geographical changes. *Carnets de Géologie/Notebooks on Geology*. Maintenon, **CG2004 A01 CCE PG**: 1-10.

Esper, O., Zonneveld, K. A. F., Höll, C., Karwath, B., Schneider, R., Vink, A., Weise-Ihlo, I.,, Willems, H., 2000. Reconstruction of palaeoceanographic conditions in the South Atlantic Ocean at the last two Terminations based on calcareous dinoflagellates. *International Journal of Earth Sciences*, **88**(4): 680-693.

Fauquette, S., Guiot, J., Suc, J.P., 1998. A method for climatic reconstruction of the Mediterranean Pliocene using pollen data. *Palaeogeography, Palaeoclimatology, Palaeoecology*, **144**: 183-201.

Friedrich, O., Meier, K.J.S., 2003. Stable isotopic indication for cyst formation depth of Campanian/Maastrichtian calcareous dinoflagellates. *Micropaleontology*, **49**(4): 375-380.

Fütterer, D., 1977. Distribution of calcareous dinoflagellates in Cenozoic sediments of Site 366, Eastern North Atlantic. *Inital Reports, Deep Sea Drilling project*, **41**: 533-541.

Gottschling, M., Plötner, J., 2004. Secondary structure models of the nuclear internal transcribed spacer regions and 5.8S rRNA in Calciodinelloideae (Peridiniaceae) and other dinoflagellates. *Nucleic Acids Research*, **32**(1): 307-315.

Gottschling, M., Keupp, H. Plötner, J., Knop, R., Willems, H., Kirsch, M., 2005. Phylogeny of calcareous dinoflagellates as inferred from IST and ribosomal sequence data. *Molecular Phylogenetics and Evolution*, **36**: 444-455.

Griffin, D.L., 1999. The late Miocene climate of northeastern Africa: unravelling the signals in the sedimentary succession. *Journal of the Geological Society*, **156**: 817-826.

Hernández-Becerril, D.U., Bravo-Sierra, E., 2004. New records of planktonic dinoflagellates (Dinophyceae) from the Mexican Pacific Ocean. *Botanica Marina*, **47**(5): 417-423.

Hildebrand-Habel, T., Willems, H., 1997. Calcareous dinoflagellate cysts from the Middle Coniacian to Upper Santonian chalk facies of Lägerdorf (N Germany). *Courier Forschungsinstitut Senckenberg*, **201**: 177-199.

Hildebrand-Habel, T., Willems, H., Versteegh, G.J.M., 1999. Variations in calcareous dinoflagellate associations from the Maastrichtian to Middle Eocene of the western South Atlantic Ocean (Sao Paulo Plateau, DSDP Leg 39, Site 356). *Review of Palaeobotany and Palynology*, **106**(1-2): 57-87.

Hildebrand-Habel, T., Willems, H., 2000. Distribution of calcareous dinoflagellates from the Maastrichtian to early Miocene of DSDP Site 357 (Rio Grande Rise, western South Atlantic Ocean). *International Journal of Earth Sciences*, **88**: 694-707.

Hildebrand-Habel, T., Streng, M., 2003. Calcareous dinoflagellate associations and Maastrichtian–Tertiary climatic change in a high latitude core (ODP Hole 689B, Maud Rise, Weddell Sea). *Palaeogeography, Palaeoclimatology, Palaeoecology*, **197**: 293–321.

Hilgen, F.J., 1991. Astronomical calibration of Gauss to Matuyama sapropels in the Mediterranean and implication for the geomagnetic polarity time scale. *Earth and Planetary Science Letter*, **104**: 226-244.

Höll, C., Zonneveld, K.A.F., Willems, H., 1998. On the ecology of calcareous dinoflagellates: The Quaternary Eastern Equatorial Atlantic. *Marine Micropaleontology*, **33**(1-2): 1-25.

Höll, C., Karwath, B., Rühlemann, C., Zonneveld, K.A.F., Willems, H., 1999. Palaeoenvironmental information gained from calcareous dinoflagellates: the late Quaternary eastern and western tropical Atlantic Ocean in comparison. *Palaeogeography, Palaeoclimatology, Palaeoecology*, **146**: 147-164.

Höll, C., Kemle-von Mücke, S., 2000. Late Quaternary upwelling variations in the eastern equatorial Atlantic Ocean as inferred from dinoflagellate cysts, planktonic foraminifera, and organic carbon content. *Quaternary Research*, **54**(1): 58-67.

Hsü, K.J., 1978. The Messinian salinity crisis. *Naturwissenschaften (Historical Archive)*, **5**(3): 151pp.

Hsü, K.J., Bernoulli, D., 1978. Genesis of the Tethys and the Mediterranean. *Initial Reports of the Deep Sea Drilling Project*, **42**: 943-9.

Hsü, K.J., Cita, M.B., Ryan, W.B.F., 1973. The origin of the Mediterranean evaporites. *Initial Reports of the Deep Sea Drilling Project*, **13**: 1203–1231.

Hsü, K.J., Montadert, L., Bernoulli, D., Cita, M.B., Erickson, A., Garrison, R.E., Kidd, R.B., Mèlierés, F., Müller, C., Wright, R., 1977. History of the Mediterranean salinity crisis. *Nature*, **267**: 399-403.

Iaccarino, S.M., Castradori, D., Cita, M.B., Di Stefano, E., Gaboardi, S., McKenzie, J.A., Spezzaferri, S., Sprovieri, R., 1999. The Miocene-Pliocene boundary and the significance of the earliest Pliocene flooding in the Mediterranean. *Memorie della Societa Geologica Italiana*, **54**: 109-131.

Janofske, D., 1996. Ultrastructure types in recent "Calcispheres". *Bulletin de l'Institut océanographique*, **14**(4): 295-303.

Janofske, D., Karwath, B., 2000. Oceanic calcareous dinoflagellates of the equatorial Atlantic Ocean: Cyst-theca relationship, taxonomy and aspects on ecology. *In*: Karwath, 2000, Ecological Studies on Living and Fossil Calcareous Dinoflagellates of the Equatorial and Tropical Atlantic Ocean. *Berichte Fachbereich Geowissenschaften*, Universität Bremen, **152**: 94-136.

Karwath, B., Janofske, D., Willems, H., 2000. Spatial distribution of the calcareous dinoflagellate *Thoracosphaera heimii* in the upper water column of the tropical and equatorial Atlantic. *International Journal of Earth Sciences*, **88**: 668-679.

Keupp, H., Kohring, R., 1993. Kalkige Dinoflagellatenzysten aus dem Obermiozän von El Medhi (Algerien). *Berliner Geowisswissenschaftliche Abhandlungen*, **9**: 25-43.

Keupp, H., Bellas, S.M., Frydas, D., Kohring, R., 1994. Aghia Irini, ein Neogenprofil auf der Halbinsel Gramvoússa/NW-Kreta. *Berliner Geowissenschaftliche Abhandlungen*, **13**(E): 469-481.

Keupp, H., Kohring, R., 1999. Kalkiger Dinoflagellatenzysten aus dem Obermiozän (NN 11) W von Rethymnon (Kreta). *Berliner Geowissenschaftliche Abhandlungen*, **30**(E): 33-53.

Keupp, H., Versteegh, G., 1989. Ein neues Konzept für kalkige Dinoflagellaten-zysten der Subfamilie Orthopithonelloideae Keupp 1987. *Berliner Geowissenschaftliche Abhandlungen*, **106**(A): 207-219.

Keogh, S.M., Butler, R.W.H., 1999. The Mediterranean water body in the late Messinian: interpreting the record from marginal basins on Sicily. *Journal of the Geological Society, London*, **156**: 837-846.

Kohring, R. 1993. Kalkdinoflagellaten aus dem Mittel- und Obereozän von Jütland (Dänemark) und dem Pariser Becken (Frankreich) im Vergleich mit anderen Tertiär-Vorkommen. *Berliner Geowissenschaftliche Abhandlungen*, **E 6**: 1-164.

Kohring, R., 1993. Kalkdinoflagellaten-Zysten aus dem unteren Pliozän von E-Sizilien. *Berliner Geowissenschaftliche Abhandlungen*, **E 9**: 15-23.

Kohring, R., 1997. Calcareous dinoflagellate cysts from the Blue Clay formation (Serravalian, Late Miocene) of the Maltese Islands. *Neues Jahrbuch, Geologisch-Paläontologische Mitteilungen*, **3**: 151-164.
Kohring, R., Gottschling, M., Keupp, H. 2005. Examples for character traits and palaeoecological significance of calcareous dinoflagellates. *Paläontologische Zeitschrift*, **79**(1): 79-91.

Kouwenhoven, T.J., 2000. *Survival under stress: benthic foraminiferal patterns and Cenozoic biotic crises, Geologica Ultraiectina*, Universiteit Utrecht (Faculteit Aardwetenschappen) **186**: 206 pp.

Kouwenhoven, T. J., Morigi, C., Negri, A., Giunta, S., Krijgsman, W., Rouchy, J. M., 2006. Paleoenvironmental evolution of the eastern Mediterranean during the Messinian: Constraints from integrated microfossil data of the Pissouri Basin (Cyprus). *Marine Micropaleontology*, **60**(1): 17-44.

Krijgsman, W., Langereis, C.G., Zachariasse, W.J., Boccaletti, M., Moratti, G., Gelati, R., Iaccarinof, S., Papani, G., Villa, G., 1999. Late Neogene evolution of the Taza–Guercif Basin (Rifian Corridor,Morocco) and implications for the Messinian salinity crisis. *Marine Geology*, **153**:147–160.

Krijgsman, W., Blanc-Valleron, M.-M., Flecker, R., Hilgen, F. J., Kouwenhoven, T. J., Merle, D., Orszag-Sperber, F., Rouchy, J. M. 2002. The onset of the Messinian salinity crisis in the eastern Mediterranean (Pissouri Basin, Cyprus). *Earth Planetary Science Letters*, **194**: 299– 300.

Logan, A., Bianchi, C.N., Morri, C., Zibrowius, H. 2004. The present-day Mediterranean brachiopod fauna: diversity, life habits, biogeography and paleobiogeography. *Scientia Marina*, **68**(1): 167-170.

Loget, N., Davy, P., Van Den Driessche, J., 2006. Mesoscale fluvial erosion parameters deduced from modelling the Mediterranean Sea level drop during the Messinian (late Miocene). *Journal of Geophysical Research*, **111**: F03005.

Lofi, C.J., Gorini, C., Berné, S., Clauzon, G., Reis, A.T.D., Ryan, W.B.F., Steckler, M.S., 2005. Erosional processes and paleo-environmental changes in the Western Gulf of Lions (SW France) during the Messinian Salinity. *Marine Geology*, **217**(1-2): 1-30

Manzi, V., Lugli, S., Lucchi, F.R., Roveri, M., 2005. Deep-water clastic evaporites deposition in the Messinian Adriatic foredeep (northern Apennines, Italy): did the Mediterranean ever dry out? *Sedimentology*, **52**: 875–902.

Matano, F., Barbieri, M., Di Nocera, S., Torre, M., 2005. Stratigraphy and strontium geochemistry of Messinian evaporite-bearing successions of the southern Apennines foredeep, Italy: implications for the Mediterranean "salinity crisis" and regional Palaeogeography. *Palaeogeography, Palaeoclimatology, Palaeoecology*, **217**(1-2): 87-114.

Meier, K.J.S., Janofske, D., Willems, H., 2002. New calcareous dinoflagellates (Calciodinelloideae) from the Mediterranean Sea. *Journal of Phycology*, **38**: 602-615.

Meier, K.J.S., Willems, H. 2003. Calcareous dinoflagellate cysts from surface sediments of the Mediterranean Sea: distribution patterns and influence of main environmental gradients. *Marine Micropaleontology*, **48**: 321-354.

Meier, K. J. S., Zonneveld, K. A. F., Kasten, S., Willems, H. 2004. Different nutrient sources forcing increased productivity during eastern Mediterranean S1 sapropel formation as reflected by calcareous dinoflagellate cysts. *Paleoceanography*, **19**: 1– 2.

Meijer, P.T., Krijgsman, W., 2005. A quantitative analysis of the desiccation and re-filling of the Mediterranean during the Messinian Salinity Crisis. *Earth and Planetary Science Letters*, **240**(2): 510-520.

Orszag-Sperber, F., Rouchy, J.M., Elion, P., 1989. The sedimentary expression of regional tectonic events during the Miocene-Pliocene transition in the southern Cyprus basins. *Geological Magazine*, **126**(3): 291-299.

Orszag-Sperber, F, Rouchy, J.-M., Blanc-Valleron, M.-M., 2000. La transition Messinien–Pliocène en Méditerranée orientale (Chypre) : la période du *Lago-Mare* et sa signification. *Earth and Planetary Sciences*, **331**: 483-490.

Robertson, A.H.F., Eaton, S.E., Folloes, E.J., Payne, A.S., 1995. Sedimentology and depositional processes of Miocene evaporites from Cyprus. *Terra Nova*, **7**: 233-254.

Rossignol-Strick, M., 1985. Mediterranean Quaternary sapropels, an immediate response of the African monsoon to variation of insolation. *Palaeogeography, Palaeoclimatology, Palaeoecology*, **49**: 237-263.

Rouchy, J.M., Orszag-Sperber, F., Blanc-Valleron, M.-M., Pierre, C., Rivière, M., Combourieu-Nebout, N., Panayides, I., 2001. Paleoenvironmental Changes at the Messinian–Pliocene Boundary in the Eastern Mediterranean (Southern Cyprus basins): Significance of the Messinian Lago –Mare. *Sedimentological Geology*, **145**: 93–117.

Seidenkrantz, M.-S., Kouwenhoven, T.J., Jorissen, F.J. Shackleton, N.J., van der Zwaan, G.J., 2000. Benthic foraminifera as indicators of changing Mediterranean–Atlantic water exchange in the late Miocene. Marine Geology, **163**: 387–407.

Spezzaferri, S., Cita, M.-B., McKenzie, J.A., 1998. The Miocene/Pliocene boundary in the eastern Mediterranean: Results from sites 967 and 969. *Proceedings of the Ocean Drilling Program, Scientific Results,* **160**: 9-28.

Steenbrink, J., Hilgen, F.J. Krijgsman, W., Wijbrans, J.R., Meulenkamp, J.E., 2006. Late Miocene to Early Pliocene depositional history of the intramontane Florina-Ptolemais-Servia Basin, NW Greece: Interplay between orbital forcing and tectonics. *Palaeogeography, Palaeoclimatology, Palaeoecology*, **238**(1-4): 151-178.

Streng, M., Hildebrand-Habel, T., Willems, H. 2002. Revision of the genera *Sphaerodinella* Keupp and Versteegh, 1989 and *Orthopithonella* Keupp *in* Keupp and Mutterlose, 1984 (Calciodinelloideae, calcareous dinoflagellate cysts). *Journal of Paleontology*, **76**: 397-407.

Streng, M., Hildebrand-Habel, T., Willems, H., 2004. Long-term evolution of calcareous dinoflagellate associations since the Late Cretaceous: comparison of a high- and a low-latitude core from the Indian Ocean. *Journal of Nannoplankton Research*, **26**: 13-45.

Tanimura, Y., Shimada, C. 2004. Calcareous dinoflagellates from a northwestern Pacific sediment trap and their paleoceanographic implications. *Micropaleontology*, **50**: 343-356.

Van der Laan, E., Snel, E., de Kaenel, E., Hilgen, F. J., Krijgsman, W., 2006. No major deglaciation across the Miocene-Pliocene boundary: Integrated stratigraphy and astronomical tuning of the Loulja sections (Bou Regreg area, NW Morocco). *Paleoceanography*, **21**: 1-27.

Versteegh G.J.M., 1993. New Pliocene and Pleistocene calcareous dinoflagellate cysts from southern Italy and Crete. *Review of Palaeobotany and Palynology*, **78**: 353-380.

Vink, A., 2004. Calcareous dinoflagellate cysts in South and equatorial Atlantic surface sediments: diversity, distribution, ecology and potential for palaeoenvironmental reconstruction. *Marine Micropaleontology*, **50**: 43-88.

Vink, A., Brune, A., Höll, C., Zonneveld, K.A.F., Willems, H., 2002. On the response of calcareous dinoflagellates to oligotrophy and stratification of the upper water column in the equatorial Atlantic Ocean. *Palaeogeography, Palaeoclimatology, Palaeoecology*,**178**: 53-66.

Vink, A., Baumann, K.-H., Boeckel, B., Esper, O., Kinkel, H., Volbers, A., Willems, H., Zonneveld, K.A.F., 2003. Coccolithophorid and dinoflagellate synecology in the South and Equatorial Atlantic: Improving the palaeoecological significance of phytoplanktonic microfossils, In: Wefer, G, Mulitza, S, Rathmeyer, V., (Eds.). *The South Atlantic in the Late Quaternary: Reconstruction of Material Budgets and Current Systems, Springer Verlag*, Heidelberg, New York: 101-20

Vink, A., Zonneveld, K.A.F., Willems, H., 2000. Distributions of calcareous dinoflagellates in surface sediments of the western equatorial Atlantic, and their potential use in palaeoceanography. *Marine Micropaleontology*, **38**: 149-180.

Wade, B.S., Bown, P.R., 2005. Calcareous nannofossils in extreme environments: The Messinian Salinity Crisis, Polemi Basin, Cyprus. *Palaeogeography, Palaeoclimatology, Palaeoecology*, **233**(3-4): 271-286.

Wehausen, R., Brumsack, H.J., 1998. The formation of Pliocene Mediterranean sapropels: Constraints from high-resolution mayor and minor element studies. *Proceedings of the Ocean Drilling Program, Scientific results*, **160**: 207-217.

Wendler, I., Zonneveld, K.A.F., Willems, H., 2002a. Calcareous cyst-producing dinoflagellates: ecology and aspects of cyst preservation in a highly productive oceanic region. *In*: Clift, P.D., Kroon, D., Geadicke, C.,, Craig, J. (Eds). The tectonic and climatic evolution of the Arabian Sea region. *Geological Society Special Publication*, **195**: 317-340.

Wendler, I., Zonneveld, K.A.F., Willems, H., 2002b. Oxygen availability effects on early diagenetic calcite dissolution in the Arabian Sea as inferred from calcareous dinoflagellate cysts. *Global and Planetary Change*, **34**: 219-239.

Wendler, I., Zonneveld, K.A.F., Willems, H,. 2002c. Production of calcareous dinoflagellate cysts in response to monsoon forcing off Somalia: a sediment trap study. *Marine Micropaleontology*, **46**: 1-11.

Ziveri, P., Rutten, A., de Lange, G., Thomson, J., Corselli, C., 2000. Present-day coccolith fluxes recorded in central Eastern Mediterranean sediment traps and surface sediments. *Palaeogeography, Palaeoclimatology, Palaeoecology*, **158**: 175-195.

Zonneveld, K.A.F., 2004. Potential use of stable oxygen isotope composition of *Thoracosphaera heimii* for upper water column (thermocline) temperature reconstruction. *Marine Micropaleontology*, **50**(3/4): 307-317.

Zonneveld, K.A.F., Brune, A., Willems, H., 2000. Spatial distribution of calcareous dinoflagellates in surface sediments of the South Atlantic Ocean between 13°N and 36°S. *Review of Palaeobotany and Palynology*, **111**: 197-223.

Zonneveld, K.A.F., Höll, C., Janofske, D., Karwath, B., Kerntopf, B., Rühlemann, C., Willems, H., 1999. Calcareous dinoflagellates as palaeo-environmental tools: 145-164p. In Fischer, G., Wefer, G. (Eds), Use of proxies in Paleoceanography: Examples from the South Atlantic. *Springer Verlag*, Berlin. 735 pp.

Zonneveld, K.A.F., Versteegh, G.J.M., De Lange, G.J., 2001. Palaeoproductivity and post-depositional aerobic organic matter decay reflected by dinoflagellate cyst assemblages of the Eastern Mediterranean S1 sapropel. *Marine Geology*, **172**: 181-195.

Chapter 3

Palaeoenvironmental changes of the early Pliocene (Zanclean) in the eastern Mediterranean Pissouri Basin (Cyprus) evidenced from calcareous dinoflagellate cyst assemblages

Katarzyna-Maria Bison[1,*], Gerard J.M. Versteegh[1], Fabienne Orszag-Sperber[3], Jean Marie Rouchy[2] and Helmut Willems[1].

[1] *Universität Bremen, Fachbereich Geowissenschaften, Postfach 330440, D-28334 Bremen, Germany.*

[2] *Muséum National d'Histoire Naturelle, 43 rue Buffon, 75005 Paris, France.*

[3] *Université de Paris-Sud, Département des Sciences de la Terre, 91405 Orsay, France*

[*]Corresponding author: kbison@uni-bremen.de

***Marine Micropaleontology* (2009), 73: 49-56**

(Received 18 December 2008, Received in revised form 24 June 2009, Accepted 25 June 2009)

Abstract

The first ~100 ka of the earliest Pliocene (Zanclean) sediments of the eastern Mediterranean Pissouri Basin on Cyprus have been investigated on calcareous dinoflagellate cyst assemblages. These assemblages reflect the return, in three phases, to the open oceanic conditions following the Messinian Salinity Crisis (MSC) 5.33 Ma ago. The lowermost phase, (dominated by *Leonella granifera* cysts) indicates substantial land-derived nutrient supply and low salinities. This confirms earlier observations of enhanced continental water runoff during the earliest Pliocene in the eastern Mediterranean basin. The second phase (dominated by *Caracomia stella* and *Calciodinellum albatrosianum* cysts) indicates a change to warmer, meso- to oligotrophic waters. The third phase marks the first appearance and still continuing dominance of *Thoracosphaera heimii* in the region. In addition, the endemic eastern Mediterranean species, *Lebessphaera urania*, peaks within this upper interval. It probably survived the Salinity Crisis in the Mediterranean. The assemblage reflects the establishment of typical open marine, well stratified and oligotrophic surface conditions similar but not yet identical to the situation in the Mediterranean Sea today. This study points out a discrepancy between the fast refilling scenario (~1 – 2 ka) in the Pissouri Basin and a long time (~100ka) environmental reorganisation in the surface waters as indicated by the calcareous dinoflagellate cysts. The onset of the Pliocene is also marked by the first appearance of *Calciodinellum elongatum* and *Calciodinellum levantinum*, which must have migrated from the Atlantic Ocean.

Keywords: Messinian Salinity Crisis, calcareous dinoflagellates, eastern Mediterranean, early Pliocene (Zanclean), Pissouri Basin (Cyprus)

Introduction

The Messinian salinity crisis (MSC) reflects a period during which the exchange between the Mediterranean and Atlantic was, at least, seasonally restricted. This caused strong changes in the Mediterranean realm and resulted in mostly hypersaline conditions during the climax of the MSC followed by brackish to freshwater environments (the so-called Lago-Mare) at the end of the crisis and again normal marine conditions at the beginning of the Pliocene (Cita *et al.*, 1978). These drastic changes had severe effects on the Mediterranean deep-water fauna (e.g. Cita, 1976; Iaccarino, 1985; Seidenkrantz *et al.*, 2000; Emig and Geistdoerfer, 2004; Kouwenhoven *et al.*, 2006). Many species disappeared from the Mediterranean Sea; others survived in marine refuges or re-entered the Mediterranean from the Atlantic Ocean upon replenishment of the Mediterranean basin (Bianchi and Morri, 2000; Logan *et al.*, 2004; Bison *et al.*, 2007). Many studies address the Miocene-Pliocene transition and the re-establishing of normal marine conditions (e.g. Hsü *et al.*, 1973; Cita *et al.*, 1978; McKenzie and Sprovieri, 1990; Fortuin *et al.*, 1995; Blanc-Valleron *et al.*, 1998; Castradori, 1998; Pierre *et al.*, 1998; Spezzaferri *et al.*, 1998; Di Stefano *et al.*, 1999; Iaccarino *et al.*, 1999; Iaccarino and Bossio, 1999; Orszag-Sperber *et al.*, 2000; Pierre *et al.*, 2006; Rouchy *et al.*, 2001, 2007). In a previous study (Bison *et al.*, 2007) we addressed the environmental changes related to the transition into the Messinian for the Pissouri Basin on Cyprus based on the evolution of its calcareous dinoflagellate cyst assemblages.

In this paper, we report the analysis of the distribution of the calcareous dinoflagellate cysts and their environmental evolution, in the first meter of the Pliocene, in order to precise the reestablishment of the marine conditions after the MSC.

Calcareous dinoflagellates

These marine planktic photosynthetic calcareous-cyst-producing organisms have been studied throughout a wide range of marine environments for their species-environment relations (e.g. Meier and Willems, 2003; Vink, 2004; Zonneveld *et al.*, 2005; Richter *et al.*, 2007). Most of the recent calcareous dinoflagellate species can be traced back at least to the Oligocene (Fütterer, 1977; Hildebrand-Habel *et al.*, 1999; Hildebrand-Habel and Willems, 2000; Hildebrand-Habel and Streng, 2003; Streng *et al.*, 2004; Kohring, 1993a) which facilitates application of the present-day species-environment relations for reconstructing past environments by means of fossil assemblages (e.g. Kohring, 1993a, Keupp *et al.*, 1994; Keupp and Kohring, 1999; Esper *et al.*, 2000; Hildebrand-Habel and Willems, 2000; Meier *et al.*, 2004; Bison *et al.*, 2007). Furthermore, most calcareous dinoflagellate cyst species appear to resist carbonate dissolution better than other calcareous plankton during transport to the sea floor and after deposition (Vink *et al.*, 2000; Baumann, *et al.*, 2003; Meier *et al.*, 2004; Zonneveld, 2004).

Geological background

The Pissouri Basin, in the Southern part of the Cyprus island, corresponds to a small tectonically depression elongated NNW-SSE (Fig. 1). In the late Miocene (Messinian) the ongoing uplift of the Troodos Massif (Fig. 1) was superimposed by sea level fall caused by the progressive restriction and partial drying out during the MSC (Hsü *et al.*, 1978). Deep water conditions were re-established with the beginning of the Pliocene (Stow *et al.*, 1995; Rouchy *et al.*, 2001; Krijgsman *et al.*, 2002; Di Stefano *et al.*, 1999). The Pliocene marls ("Trubi") of the Pissouri Basin conformably overly the upper most Messinian (between 5.59 and 5.50 Ma) "Lago Mare" sediments, consisting of a complex succession of marls, gypsum, carbonates, conglomerates, intercalated with palaeosols and indicating periods of emersion at the end of the Messinian, with a sharp sedimentary contact (Orszag-Sperber *et al.*, 2000; Rouchy *et al.*, 2001). Such sharp sedimentary contact at the Miocene/Pliocene boundary can be observed Mediterranean-wide and has been generally associated with fast refilling of the sedimentary basins (e.g. Rouchy *et al.*, 2001; Orszag-Sperber *et al.*, 2000; Meijer and Krijgsman, 2005; Loget and Van Den Driessche, 2006; Orszag-Sperber, 2006).

Climate

The Mediterranean climate was dry with some humid phases during most of the Messinian (Griffin, 2002). The late Messinian Lago Mare period follows the deep desiccation of the Mediterranean basin (Krijgsman *et al.*, 1999). It probably reflects a climatic shift, associated with increased precipitation and runoff throughout the Mediterranean area including the northern borderlands (Rizzini and Dondi, 1979; Willett *et al.*, 2006). It also became more humid off north west Africa (Ruddiman *et al.*, 1989), in the Arabian Sea (De Menocal and Bloemendal, 1995), the Gulf of Suez/Red Sea (Griffin, 1999) and the central European Alps (Willett *et al.*, 2006). Oxygen isotope records of benthic foraminifera in the Atlantic Ocean (Hodell *et al.*, 2001; Vidal *et al.*, 2002) indicate that expansion of the global ice volume ended at precisely 5.55 Ma coincident with the onset of the Lago Mare conditions. This event resulted in a warming of the Atlantic Ocean (Willett *et al.*, 2006) intensifying precipitation in Europe and consequently changing the water budget for the Mediterranean (Willett *et al.*, 2006). Apparently, this latest Miocene period with a warm and wet postglacial climate persisted into the early Pliocene (Willett *et al.*, 2006).

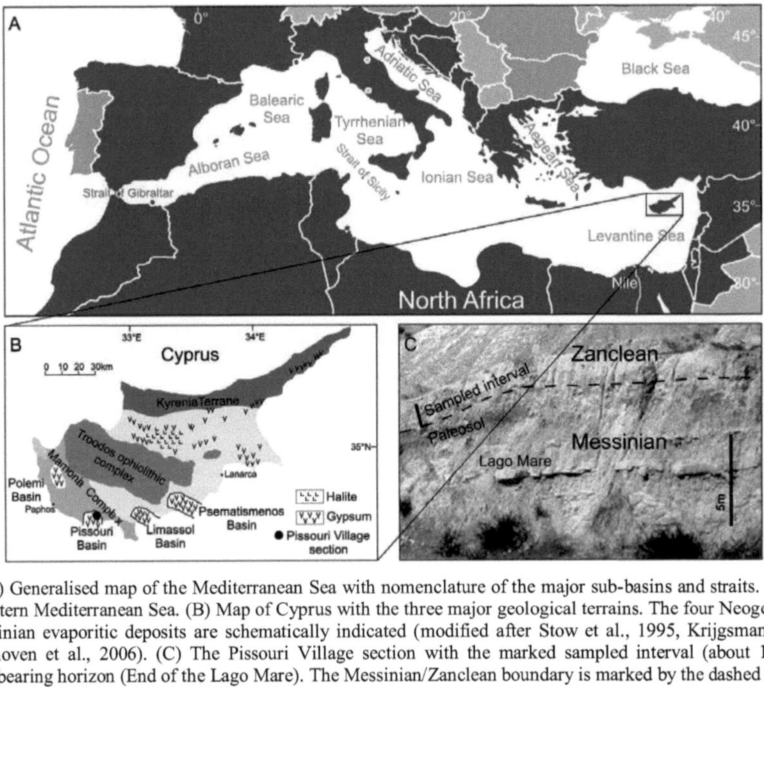

Fig. 1 (A) Generalised map of the Mediterranean Sea with nomenclature of the major sub-basins and straits. Cyprus is located in the eastern Mediterranean Sea. (B) Map of Cyprus with the three major geological terrains. The four Neogene subbasins and the Messinian evaporitic deposits are schematically indicated (modified after Stow et al., 1995, Krijgsman et al., 2002 and Kouwenhoven et al., 2006). (C) The Pissouri Village section with the marked sampled interval (about 1.20m) above the paleosol-bearing horizon (End of the Lago Mare). The Messinian/Zanclean boundary is marked by the dashed line.

Material

The Pissouri section outcrops along the old Limassol-Paphos road (Fig. 1). This section was originally described by Orszag-Sperber and Rouchy (1979), then by Rouchy and Pierre (1979). Di Stefano *et al.* (1999) re-sampled this section in order to precise the chronology of the earlier Pliocene.The samples were collected at the base of the Pliocene (already dated by Bizon, in Orszag-Sperber and Rouchy, 1979; then detailed by Di Stefano *et al.*, 1999), up to about 12 meters, every 10 centimetres. A total of 14 marls from the earliest Pliocene part of the Pissouri Village section (see Fig. 2) were investigated. Based on biostratigraphical studies (Di Stefano *et al.*, 1999) the investigated Pliocene interval spans about 100 ka. The first 8 cm of these early Pliocene sediments consist of foraminifer rich variegated marls mainly composed of epipelagic assemblages (*Globigerinoides trilobus*, *Globigerinoides obliquus* and *Globigerinoides sacculifer*) indicating shallow marine conditions. However, the lack of stratigraphic markers does not allow attributing a precise age to these marls. The subsequent beige and white marls display a typical Trubi facies still contain epipelagic microfauna at the base but benthic microfauna appears at 14 cm (Rouchy *et al.*, 2001). The base of this interval belongs to the *Sphaerodinellopsis* acme zone (Bizon, in Orszag-Sperber and Rouchy, 1979; MPL1, Di Stefano *et al.*, 1999). The planktic foraminifera indicate a water depth of at least 300 m (Di Stefano, 1999).

Methods

For each sample ~ 0.5 g of dried sediment was disaggregated in water, and the 20-75 μm fraction was isolated by sieving. An aliquot of these fractions was mounted on a stub for quantitative and qualitative analysis using a Scanning Electron Microscope (Bison *et al.*, 2007). Additionally, selected samples were prepared for polarised light microscopic investigations with a Gypsum plate (Janofske, 1996) to identify the crystallographic orientation of the wall crystals which is an additional character to identify the cysts at the genus level (Janofske, 2000; Janofske and Karwath, 2000). Calcareous dinoflagellate cysts were absent in the 5-20 μm and >75 μm fractions of randomly selected pilot samples. For diversity calculations the Shannon-Weaver index was used.

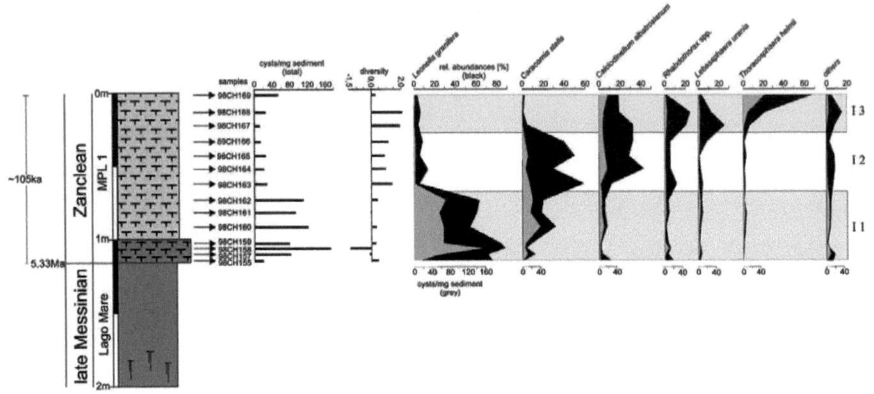

Fig. 2 Stratigraphic profile of the Messinian/Zanclean transition of the Pissouri Village section (after Orszag-Sperber et al., 2000; Rouchy et al., 2001) and positions of samples studied. Age, diversity and cysts abundances in cysts per mg sediment of the analysed samples are displayed with relative abundances of *L. granifera*, *C. stella*, *C. albatrosianum*, *Rhabdothorax spp.*, *L. urania*, *T. heimii* and other cysts against time. I1 = *Leonella granifera* interval; I2 = *Caracomia stella/Calciodinellum albatrosianum* interval; I3 = *Thoracosphaera heimii/C. albatrosianum* interval.

Results

We examined both relative and absolute abundances (cysts mg^{-1} of dry sediment) of calcareous dinoflagellate cysts in 14 samples collected in the earliest Pliocene Pissouri Basin which were all productive (Fig. 2). Three intervals have been differentiated based on the shifts in calcareous dinoflagellate cyst dominance (Fig. 2).

The lowermost "*Leonella granifera* interval" has the highest cyst concentrations (Fig. 2). It is dominated by *L. granifera* (average of 70%). After its prominent peak (both in relative and absolute abundances), *Caracomia stella* becomes more abundant (Fig. 2).

The transition to the second "*C. stella/Calciodinellum albatrosianum* interval" is characterised by a significant drop in overall cyst concentrations (Fig. 2). *L. granifera* concentrations decrease rapidly, those of *C. stella* more gradually. *C. albatrosianum* slightly increases. The instant drop in *L. granifera* causes both *C. stella* and *C. albatrosianum*, to reach maximum relative abundances and induces a slight peak in relative abundance of *Rhabdothorax* spp.

The third and upper "*Thoracosphaera heimii/C. albatrosianum* interval" is characterised by a strong drop of *C. stella* whereas *T. heimii* rises to dominance and *C. albatrosianum* becomes abundant (Fig. 2). *Lebessphaera urania* and *Rhabdothorax* spp. are also well represented. Towards the end, cyst concentrations increase again due to the rise of *T. heimii*.

Discussion

With the refilling of the Mediterranean basin during the earliest Pliocene, *Lebessphaera urania*, a postulated remnant of the Tethyan Ocean (Meier and Willems, 2003), reappears but assemblages dominated by *C. albatrosianum* during Tortonian/Messinian time (Bison *et al.*, 2007) are missing. Instead, a succession of calcareous dinoflagellate cyst assemblages is observed, first dominated by *L. granifera*, then by *C. stella* and *C. albatrosianum* and finally by *T. heimii* and *C. albatrosianum*. Furthermore, the quantitatively less important Atlantic species *C. elongatum* (Bison *et al.*, 2007) and *C. levantinum* have their first appearance. The environmental evolution indicated by this succession will be discussed below.

Leonella granifera interval

L. granifera appears to be strongly connected to elevated nutrient/mineral availability, in particular as they are transported by rivers (Wendler *et al.*, 2002a, b, c; Vink et al, 2002; Vink, 2004) and/or winds associated with upwelling of nutrient-rich waters such as at the coast of NW Africa (Richter *et al.*, 2007) or at the Benguela upwelling area of SW Africa (Vink, 2004). Intrinsically, these conditions have an unstable character and often a strong seasonal component. In the Eastern Mediterranean the species is very rare throughout the Holocene and at present (Meier and Willems, 2003) except during the formation of the organic-rich S1 sapropel (Meier *et al.*, 2004). The formation of sapropel S1 correlates positively with a strong terrigenous input (Bianchi *et al.*, 2006) and reduced salinity due to increased freshwater runoff, predominantly and seasonally from the Nile River (e.g. Rossignol-Strick, 1985, 1987; Hilgen, 1991; Rohling and Hilgen, 1991; Meier *et al.*, 2004, Meijer and Tuenter, 2007). This suggests that, in the past Eastern Mediterranean, the presence of *L. granifera* could indicate an eutrophication of environments subject to direct terrestrial influx primarily by river discharge.

Conversely we assume that also in the Pissouri Basin it indicates these conditions at least seasonally. These may have been in combination with local upwelling caused by an estuarine circulation pattern, as it is shown by van Harten (1984) for the eastern Mediterranean basin (Crete) during the early Pliocene. After van Harten (1984), the estuarine circulation, affecting both the eastern and western Mediterranean basin of the early Pliocene, was strongly associated with increased atmospheric humidity and driven by upwelling.

This interpretation agrees with the lighter stable oxygen isotope composition in the Pissouri Basin, during (~-5‰) and directly after (~-3‰) the Lago Mare period, during the earliest Pliocene (Rouchy et al, 2001, 2007). The shift from lighter to heavier oxygen isotope values is much more transitional in the Pissouri Basin than offshore (Rouchy *et al.*, 2001), confirming its enhanced continental interference. It also supports the proposed continuation of increased fluvial input in the eastern Mediterranean basin after the latest Messinian brackish-freshwater period (e.g. Diester-Haass *et al.*, 1998; Di Stefano *et al.*, 1999, Iaccarino *et al.*, 1999; Rouchy *et al.*, 2001, 2007; Willet *et al.*, 2006).

C. stella/C. albatrosianum interval

The change to dominance of *C. stella* and *C. albatrosianum* is primarily passive and driven by the decrease of *L. granifera*. This implies that the influence of the probably strongly seasonal riverine terrestrial input and eutrophic conditions considerably decreased resulting in a replacement by the formerly background signal, as it is indicated by *L. granifera*. Fossil *C. stella* has so far only been described from warmer environments of low and middle latitudes inside the Mediterranean (Keupp and Versteegh, 1989; Kohring, 1993a, b, 1997; Keupp and Kohring, 1999; Bison *et al.*, 2007) and outside (Fütterer, 1977, as *Thoracosphaera* sp 2; Hildebrand-Habel and Willems, 2000; Streng *et al.*, 2002, 2004). Its occurrence in a trap sample from the western tropical Atlantic Ocean as *Thoracosphaera albatrosiana* "granular form" sensu Fütterer 1977 (Dale, 1992) agrees with the (fossil)-inferred preference for warm mesotrophic and more shelfward environments (Keupp and Kohring, 1999; Streng *et al.*, 2002, 2004; Bison *et al.*, 2007) with normal or even slightly increased salinities. These conditions thus replace the environmental conditions during the lowermost phase. Since the relative raise of *C. stella* coincides with a decrease in its concentrations (Fig. 2), we suggest, that after the collapse of *L. granifera,* also for *C. stella* the environmental conditions become less favourable. The much more drastic change for *L. granifera* probably relates to its habit to shed shells much more often (during each vegetative reproduction (Meier *et al.*, 2007)) than *C. stella* (sexual reproduction only) so that relative to other calcareous dinoflagellates its sedimentary abundance is much more sensitive to population changes in the surface waters. Apart from niche overlap, temporal (seasonal) variability probably contributed to the co-occurrence of *C. stella* and *L. granifera* in our samples.

C. albatrosianum, (the second most abundant species of this interval) today dominates the calcareous-cyst assemblages in open marine environments with relatively warm (sub)surface waters and a low seasonal temperature amplitude (e.g. Janofske and Karwath, 2000; Zonneveld *et al.*, 2000; Wendler *et al.*, 2002b, c; Vink *et al.*, 2000; Vink, 2004; Richter *et al.*, 2007). It is also abundant in highly productive and nutrient-rich regions (Wendler *et al.*, 2002b; Richter *et al.*, 2007), suggesting an adaptation to a wide range of nutrient levels (Vink, 2004). In sediments below the oligotrophic gyres of the Worlds' Oceans *C. albatrosianum* is outnumbered by *T. heimii*. This may not represent the situation in the water column since, like *L. granifera, T. heimii* produces vegetative cysts and is thus overrepresented in the underlying sediments. *C. albatrosianum* is less abundant in coastal areas (Vink, 2004; Wendler *et al.*, 2002b, c; Richter *et al.*, 2007). Thus, high abundances of this species reflect relatively stable and warm open oceanic environments with low seasonal temperature variations and normal salinities, whereby high temperatures are most important (Vink, 2004). *C. albatrosianum* is notably absent in the Mediterranean Sea today, which is attributed to the Mediterranean high annual temperature amplitude (Vink, 2004).

The presence of *C. albatrosianum* thus further supports a change from a highly productive, strongly terrestrially influenced period with variable salinity towards more stable, meso- to oligotrophic marine conditions and prescribes a warm climate. However, compared to the assemblages prior to the MSC (Bison *et al.*, 2007), relative abundances of *C. albatrosianum* decreased, which may indicate a relative (seasonal) cooling.

The observed environmental change reflects a general shift in the climate situation since the late Messinian from relatively humid to more dry conditions in the eastern Mediterranean basin and central and eastern Asia (Fortelius *et al.*, 2006). The increasing evaporation accompanied by progressing intrusion of Atlantic waters into the eastern Mediterranean basin must have led to a rise in salinity and probably the establishment of an anti-estuarine circulation pattern comparable to the present one.

Thoracosphaera heimii/Calciodinellum albatrosianum **interval**

The environmental interpretation of this interval is again mainly based on the ecology of the dominant species which are *T. heimii* and *C. albatrosianum*.

T. heimii is the most abundant calcareous dinoflagellate species in modern oceans (e.g. Dale, 1992; Dale and Dale, 1992; Zonneveld *et al.*, 2000; Karwath *et al.*, 2000; Wendler *et al.*, 2002 a, b, c; Vink, 2004; Tanimura and Shimada, 2004; Richter *et al.*, 2007) as well as in the present-day Mediterranean Sea (Meier and Willems, 2003). The start of this dominance in the World's Oceans could not be detected so far. We appear to record this start in the Mediterranean during the earliest Pliocene. *T. heimii* is characterised by a large tolerance to varying temperatures (e.g. Karwath *et al.*, 2000; Vink, 2004; Zonneveld *et al.*, 2005) with highest cyst production at lower temperatures around 16°C (Karwath *et al.*, 2000). It dominates in regions with a well developed deep chlorophyll maximum (Vink, 2004).

In the uppermost interval *C. albatrosianum* is second in abundance and *Rhabdothorax* spp. and *L. urania* are of minor abundance (Fig. 2). To a large extent this event marks the transition to the highly oligotrophic well stratified surface waters prevailing in the modern eastern Mediterranean Sea. At present, *T. heimii* vastly exceeds the abundance of the other species by a factor of about 100, whereas *L. urania* is second in abundance (Meier and Willems, 2003). In stead, in our samples *C. albatrosianum* is second in abundance, whereas it is very rare in the Mediterranean Sea today (Meier and Willems, 2003). Possibly, temperatures get too low in winter to support *C. albatrosianum* in the present-day Mediterranean. Maybe also salinities did not reach modern levels yet. This would explain why *L. urania*, a supposedly halophytic species (Meier and Willems, 2003; Meier *et al.*, 2004) from the Mediterranean and Indio-Pacific Ocean, is not abundant yet in our samples. Interestingly, this species is not yet known from the Atlantic Ocean and may have survived the salinity crisis in the Mediterranean basin itself (Bison *et al.*, 2007).

Palaeoenvironmental scenarios

The refilling at the beginning of the Zanclean is considered to have been very rapid: less than 10 000 years (Hsü *et al.*, 1973), catastrophic deluge (Cita *et al.*, 1978), instantaneous (Hilgen and Langereis, 1988), 36 years (Blanc, 2002). During the Lago Mare period the eastern Mediterranean experienced enhanced continental water supply (e.g. Griffin, 1999, 2002; Willet *et al.*, 2006; Gladstone *et al.*, 2007, among others). The presence of brackish fauna in the very early Pliocene is documented by authors (e.g. Corradini and Biffi, 1988; Iaccarino *et al.*, 1999, among others) and considered to be either the re-working of fauna (Di Stefano *et al.*, 1999) or due to the consequence of the mixing of faunas (Iaccarino *et al.*, 1999).

The presence of the *L. granifera*-phase also indicates the persistence of continental conditions into the earliest Pliocene and agrees with other evidences, such as Diester-Haass, *et al.* (1998), Iaccarino *et al.* (1999), Rouchy *et al.* (2001) and Willett *et al.* (2006). We show that the pluvial phase continues for more than 1/3 of the investigated ~100 ka interval and that it is followed by a transitional phase of similar extension (Fig. 2). Assuming a relatively stable sedimentation rate within the investigated interval, the first period lasted ~42 ka and the transitional phase (interval 2) about 35 ka. This is much longer than the up to ~12 ka lasting warm and humid Mediterranean phases associated to precession minima (Hilgen, 1991). In the Eastern Mediterranean, extreme minima in precession normally induce the formation of organic-rich layers, the so-called sapropels (e.g. Rossignol-Strick, 1985; Diester-Haass *et al.*, 1998; Emeis *et al.*, 2000; Bianchi *et al.*, 2006). Also during the earliest Pliocene such sapropels formed and indicated a deep-water stagnation (Rossignol-Strick, 1985; Hilgen, 1991; Castradori, 1998; Spezzaferri *et al.*, 1998; Steenbrink *et al.*, 2006; Van der Laan *et al.*, 2006). Albeit the Pissouri Basin was too shallow (Bison *et al.*, 2007) and well ventilated (Rouchy *et al.*, 2001), preventing sapropel formation, it records the enhanced fresh-water input, although for a much longer period than can be accounted for by a precession minimum only. Another mechanism than precession must thus have led to the continuation of the transition to modern conditions as recorded by the calcareous dinoflagellate cysts. To what extent may the presence of less dense waters (Lago Mare) in the eastern Mediterranean basin interfered or delayed the establishment of the modern-type anti-estuarine Mediterranean system? The refilling of the Mediterranean basin at the end of the MSC probably induced an estuarine circulation pushing the less dense fresh waters (Lago Mare) already present upwards (upwelling), where it mixed with the lower saline surface waters and subsequently flow out of the basin towards the Atlantic Ocean (Van Harten, 1984). Although this also could ventilate the deeper basin and suppress sapropel formation, the presence of a sapropel suggests it didn't. This interplay is not well understood, as is the duration and rate of reorganisation of the eastern Mediterranean basin during the earliest Pliocene. Future researches are needed, notably with respect to relative short timing of precession and refilling and their interference (inhibition/enhancement). Why did the low-salinity phase extend so far into the earliest Pliocene? What was the cause for the abrupt drop of *L. granifera*? Was it caused by local river load changes facilitating an anti-estuarine circulation system coupled with a nutrient loss? What was the major source of the nutrients enriched during the earliest Pliocene? Was it continental runoff, upwelling or a combination of both?

Anyway, the study of the calcareous dinoflagellate cyst assemblages allows to differentiate three intervals, at the beginning of the Zanclean, for which it is necessary to assess the duration.

Conclusions

Significant changes in calcareous dinoflagellate cyst associations within the Pissouri Basin during the Early Pliocene reflect the progressive reestablishment of normal marine conditions in three phases, in the eastern Mediterranean, just after the MSC.

- In the earliest, *L. granifera* dominated phase, the environment suggests a period of sapropel formation. Conditions were warm and eutrophic with increased continental runoff and reduced surface water salinity.

- During the transitional phase, conditions change to less restricted more stable, normal saline and meso- to oligotrophic environments. Taxa indicative of open marine and warm water conditions such as *C. albatrosianum* and *C. stella* occur.

- The third phase - dominated by *T. heimii* and *C. albatrosianum* - denotes the establishment of typical open marine, well stratified and oligotrophic surface conditions similar but not yet identical to the situation today. Winters were probably still warmer and salinities less high. *T. heimii* for the first time dominates the calcareous dinoflagellate assemblages of the Mediterranean Sea.

The MSC did not lead to a permanent removal of oceanic calcareous dinoflagellate taxa from the Pissouri Basin but resulted in the distinct replacement of the Miocene *C. albatrosianum* dominated assemblages by post-Miocene *T. heimii* ones. Furthermore it led to the introduction of *C. elongatum* and *C. levantinum* which are interpreted as Atlantic newcomers. The pluvial period, which already started with the Messinian Lago Mare period, apparently persisted up into the very early Pliocene. Herewith–these results confirm earlier observations of continuing enhanced continental water runoff into the earliest Pliocene eastern Mediterranean basin.

The most striking result of this study is to show the evidence of a discrepancy between the fast refilling scenario (~1 – 2 ka) of the Mediterranean Basin and the long time (~100 ka) reorganisation of the surface waters as indicated by the calcareous dinoflagellate assemblages.

Acknowledgements

The authors thank two anonymous reviewers for critical reading of the manuscript and for helpful suggestions. Financial support by the German Research Foundation (ProjectWI-725/19-1/2) is gratefully acknowledged.

Appendix A

The annotated listing of calcareous dinoflagellate taxa found in the investigated material follows the nomenclature recently proposed by Elbrächter et al., 2008.

Division **Dinoflagellata** (Bütschli, 1885) Fensome *et al.*, 1993

Subdivision **Dinokaryota** Fensome *et al.*, 1993

Class **Dinophyceae** Pascher, 1914

Subclass **Peridiniphycidae** Fensome *et al.*, 1993

Order **Peridiniales** Haeckel, 1894

Suborder **Peridiniineae** Autonym

Family **Thoracosphaeraceae** (Schiller, 1930) Elbrächter *et al.* 2008

Calciodinellum albatrosianum (Kamptner, 1963) Janofske and Karwath, 2000

Calciodinellum elongatum (Hildebrand-Habel *et al.*, 1999) Meier *et al.*, 2002

Calciodinellum operosum Deflandre, 1947

Calciodinellum levantinum Meier et al., 2002

Calciodinellum limbatum (Deflandre, 1948) Kohring, 1993

Calcigonellum infula (Deflandre, 1948) Keupp, 1984

Caracomia stella (Gilbert and Clark, 1983) Streng *et al.*, 2002

Lebessphaera urania Meier *et al.*, 2002

Leonella granifera (Fütterer, 1977) Janofske and Karwath, 2000

Melodomuncula berlinensis Versteegh, 1993

Pirumella parva (Bolli, 1974) Lentin and Williams, 1993

Praecalcigonellum schizosaeptum Versteegh, 1993

Rhabdothorax spp. (Kamptner, 1937) Kamptner, 1958: includes *Scrippsiella regalis* (Gaardner, 1954) Janofske, 2000 and *Scrippsiella trochoidea* (von Stein, 1983) Loeblich III, 1965

Thoracosphaera heimii (Lohmann, 1920) Kamptner, 1944

Appendix B

Count data and calculated diversity data for all analysed samples. 1 = *Calciodinellum albatrosianum*, 2 = *Calciodinellum operosum*, 3 = *Calciodinellum elongatum*, 4 = *Calciodinellum limbatum*, 5 = *Calcigonellum infula*, 6 = *Caracomia stella*, 7 = *Lebessphaera urania*, 8 = *Leonella granifera*, 9 = *Melodomuncula berlinensis*, 10 = *Calciodinellum levantinum*, 11 = *Pirumella parva*, 12 = *Praecalcigonellum schizosaeptum*, 13 = *Rhabdothorax* spp., 14 = *Thoracosphaera heimii*.

Sample	1		2		3		4		5		6		7		8		9		10		11		12		13		14		Unidentified		Total no of dinos	Diversity
	No	%	No	%	No	%	No	%	No	%	No	%	No	%	No	%	No	%	No	%	No	%	No	%	No	%	No	%	No	%		
98CH169	40	18	-	-	5	2	-	-	-	-	4	2	6	3	7	3	-	-	5	2	-	-	-	-	3	1	147	67	1	0,5	218	0.27
98CH168	30	19	2	1	1	1	-	-	1	1	2	1	18	12	9	6	1	1	3	2	10	6	3	2	37	24	36	23	3	2	156	2,01
98CH167	30	32	3	3	1	1	-	-	-	-	4	4	24	25	4	4	2	2	-	-	2	2	-	-	19	20	6	6	-	-	95	1,85
98CH166	46	33	2	1	-	-	-	-	-	-	56	41	12	9	11	8	-	-	-	-	2	1	-	-	7	5	2	1	-	-	138	1,05
98CH165	43	28	2	1	3	2	-	-	-	-	76	50	3	2	8	5	2	1	1	1	-	-	-	-	8	5	4	3	2	1	152	0.89
98CH164	73	42	5	3	2	1	-	-	-	-	50	29	4	2	23	13	-	-	2	1	1	1	-	-	9	5	2	1	2	1	173	0.91
98CH163	13	12	2	2	4	4	-	-	-	-	66	58	5	4	6	5	-	-	-	-	1	1	-	-	12	11	2	2	2	2	113	1,35
98CH162	12	5	-	-	6	3	-	-	-	-	45	20	8	3	143	62	-	-	2	1	1	0,4	-	-	7	3	3	1	2	1	229	0,39
98CH161	20	5	1	0,3	16	4	-	-	-	-	70	18	17	4	229	60	-	-	-	-	-	-	1	0,3	16	4	12	3	1	0,3	383	-0.01
98CH160	18	7	1	0,4	7	3	-	-	-	-	79	31	4	2	133	53	-	-	1	0,4	-	-	-	-	3	1	4	2	1	0,4	251	0,30
98CH159	4	2	-	-	-	-	-	-	-	-	14	9	2	1	133	82	-	-	-	-	1	1	-	-	5	3	2	1	1	1	162	0.26
98CH158	7	1	1	0,1	11	2	-	-	-	-	18	3	11	2	626	87	2	0,3	1	0,1	-	-	-	-	38	5	-	-	5	1	720	-1.37
98CH157	10	3	1	0,3	21	6	-	-	-	-	41	12	10	3	237	70	2	1	4	1	1	0,3	-	-	9	3	1	0,3	-	-	337	-0,14
98CH155	15	10	2	1	1	1	1	1	-	-	2	1	1	1	115	75	-	-	-	-	3	2	-	-	11	7	-	-	3	2	154	0.44
																														sum:	3281	

References

Baumann, K.-H., Böckel, B., Donner, B., Gerhardt, S., Henrich, R., Vink, A., Volbers, A, Willems, H., Zonneveld, K.A.F., 2003. Contribution of calcareous plankton groups to the carbonate budget of South Atlantic surface sediments. In: Wefer, G., Mulitza, S., Ratmeyer, V. (Eds.). The South Atlantic in the Late Quaternary: reconstruction of material budget and current systems. *Springer Verlag*, Berlin, Heidelberg: 81-99.

Bianchi, C.N., Morri, C., 2000. Marine biodiversity of the Mediterranean Sea: situation, problems and prospects for the future research. *Marine Pollution Bulletin*, **40**(5): 367-376.

Bianchi, D., Zavatarelli, M., Pinardi, N., Capozzi, R., Capotondi, L., Corselli, C., Masina, S., 2006. Simulations of ecosystem response during the sapropel S1 deposition event. *Palaeogeography, Palaeoclimatology, Palaeoecology*, **235**: 265-287.

Bison, K.-M., Versteegh, G.J.M., Hilgen, F.J., Willems, H., 2007. Calcareous dinoflagellate turnover in relation to the Messinian salinity crisis in the eastern Mediterranean Pissouri Basin, Cyprus. *Journal of Micropalaeontology*, **26**: 103-116.

Blanc, P.-L., 2000. Of sills and straits: a quantitative assessment of the Messinian Salinity Crisis. *Deep-Sea Research*, **47**: 1429–1460.

Blanc-Valleron, M.-M., Rouchy, J.-M., Pierre, C., Badaut-Trauth, D., Schuler, M., 1998. Evidence of Messinian non-marine deposition at site 968 (Cyprus Lower Slope). *Proceedings of the Ocean Drilling Program, Scientific Results*, College Station, TX, **160**: 437-445.

Castradori, D., 1998. Calcareous nannofossil in the basal Zanclean of the eastern Mediterranean Sea: remarks on paleoceanography and sapropel formation. *Proceedings of the Ocean Drilling Program, Scientific Results*, **160**: 113-123.

Cita, M.B., 1976. Biodynamic effects of the Messinian Salinity Crisis on the evolution of planktonic foraminifers in the Mediterranean. *Palaeogeography, Palaeoclimatology, Palaeoecology*, **20**: 23–42.

Cita, M.B., Wright, R.C., Ryan, W.B.F., Longinelli, A., 1978. Messinian paleoenvironments. In: Hsü, K.J., Montadert, L., *et al.*, (Eds.). *Initial Reports, Deep Sea Drilling Project*, **42**(1): 1003-1035.

Corradini, D., Biffi, U., 1988. Étude des dinokystes à la limite Messinien-Pliocène dans la coupe Cava Serredi, Toscane, Italie. *Bulletin des Centres de Recherches Exploration-Production Elf-Aquitaine*, **12**: 221–236.

Dale, B., 1992. Thoracosphaerids: pelagic fluxes. In: Dale, B., Dale, A.L. (Eds.). Dinoflagellate Contributions to the Deep Sea. In: *Ocean Biocoenosis Series*, Woods Hole Oceanographic Institution, **5**: 33-41.

Dale, A.L., Dale, B., 1992. Dinoflagellate contributions to the open ocean sediment flux. In: Dale, B., Dale, A.L. (Eds.). Dinoflagellate Contributions to the Deep Sea. In: *Ocean Biocoenosis Series*, Woods Hole Oceanographic Institution, **5**: 45–73

De Menocal, P.B., Bloemendal, J., 1995. Plio-Pleistocene climatic variability in subtropical Africa and the paleoenvironment of hominid evolution: a combined data model approach. In: Vrba, E.S., Denton, G.H., Partridge, T.C., Burckle, L.H. (Eds.). Paleoclimate and Evolution, with Emphasis of Human Origins. In: *Yale University Press*, New Haven: 262–288.

Di Stefano, E., Cita, M.B., Spezzaferri, S., Sprovieri, R., 1999. The Messinian–Zanclean Pissouri section (Cyprus, Eastern Mediterranean). In: Cita, M.B., McKenzie, J. (Eds.). Cycles, events, sea levels in Messinian times. *Memorie della Società Geologica Italiana*, **54:** 133–144.

Diester-Haass, L., Robert, C., Chamley, H., 1998. Paleoproductivity and climate variations during sapropel deposition in the Eastern Mediterranean Sea. In Robertson, A.H.F., Emeis, K.-C., Richter, C., Camerlenghi, A. (Eds.). *Proceedings of the Ocean Drilling Program, Scientific Results*, College Station, TX, **160**: 227–248.

Elbrächter, M., Gottschling, M., Hildebrand-Habel, T., Keupp, H., Kohring, R., Lewis, J., Meier, K.J.S., Montresor, M., Streng, M., Versteegh, G.J.M., Willems, H., Zonneveld, K., 2008. Establishing an Agenda for Calcareous Dinoflagellates (Thoracosphaeraceae, Dinophyceae) including a nomenclatural synopsis of generic names. *Taxon*, **57** (4): 1289-1303.

Emeis, K.-C., Sakamoto, T., Wehausen, R., Brumsack, H.-J., 2000. The sapropel record of the Eastern Mediterranean Sea – results of Ocean Drilling Program Leg 160. *Palaeogeography, Palaeoclimatology, Palaeoecology*, **158**: 371–395.

Emig, C., Geistdoerfer, P., 2004. The Mediterranean deep-sea fauna: historical evolution, bathymetric variations and geographical changes. *Carnets de Géologie / Notebooks on Geology. Maintenon*, CG2004 A01 CCE PG: 1-10.

Esper, O., Zonneveld, K.A.F., Höll, C., Karwath, B., Schneider, R., Vink, A., Weise-Ihlo, I., Willems, H., 2000. Reconstruction of palaeoceanographic conditions in the South Atlantic Ocean at the last two Terminations based on calcareous dinoflagellates. *International Journal of Earth Sciences*, **88**(4): 680-693.

Fortelius, M., Eronen, J., Liu, L., Pushkina, D., Tesakov, A., Vislobokova, I., Zhang, Z., 2006. Late Miocene and Pliocene large land mammals and climatic changes in Eurasia. *Palaeogeography, Palaeoclimatology, Palaeoecology*, **238**(1-4): 219pp.

Fortuin, A.R., Kelling, J.M.D., Roep, Th.B., 1995. The enigmatic Messinian–Pliocene section of Cuevas del Almanzora (Vera Basin, SE Spain) revisited-erosional features and strontium isotope ages. *Sedimentological Geology*, **97**: 177–201.

Fütterer, D., 1977. Distribution of calcareous dinoflagellates in Cenozoic sediments of Site 366, Eastern North Atlantic. *Inital Reports of the Deep Sea Drilling Project*, **41**: 709–737.

Gladstone, R., Flecker, R., Valdes, P., Lunt, D., Markwick, P., 2007. The Mediterranean hydrologic budget from a Late Miocene global climate simulation. *Palaeogeography, Palaeoclimatology, Palaeoecology,* **251**: 254-267.

Griffin, D.L., 1999. The late Miocene climate of northeastern Africa: unravelling the signals in the sedimentary succession. *Journal of the Geological Society of London*, **156:** 817-826.

Griffin, D.L., 2002. Aridity and humidity: two aspects of the late Miocene climate of North Africa and the Mediterranean. *Palaeogeography, Palaeoclimatology, Palaeoecology*, **182**(1-2): 65-91.

Hildebrand-Habel, T., Streng, M., 2003. Calcareous dinoflagellate associations and Maastrichtian–Tertiary climatic change in a high latitude core (Ocean Drilling Program Hole 689B, Maud Rise, Weddell Sea). *Palaeogeography, Palaeoclimatology, Palaeoecology*, **197**: 293–321.

Hildebrand-Habel, T., Willems, H., 2000. Distribution of calcareous dinoflagellates from the Maastrichtian to early Miocene of DSDP Site 357 (Rio Grande Rise, western South Atlantic Ocean*). International Journal of Earth Sciences*, **88**: 694-707.

Hildebrand-Habel, T., Willems, H., Versteegh, G.J.M., 1999. Variations in calcareous dinoflagellate associations from the Maastrichtian to Middle Eocene of the western South Atlantic Ocean (Sao Paulo Plateau, Deep Sea Drilling Project Leg 39, Site 356*). Review of Palaeobotany and Palynology*, **106**(1-2): 57-87.

Hilgen, F.J., Langereis, C.G., 1988. The age of the Miocene-Pliocene boundary in the Capo Rossello area (Sicily*). Earth and Planetary Science Letters*, **91**: 214-222.

Hilgen, F.J., 1991. Extension of the astronomically calibrated (polarity) time scale to the Miocene/Pliocene boundary. *Earth Planetary Science Letters*, **107**: 349–368.

Hsü, K.J., Ryan, W.B.F., Cita, M.B., 1973. Late Miocene Desiccation of the Mediterranean. *Nature*, **242**: 240-244.

Hsü, K.J., Stoffers, P., Ross, D.A., 1978. Messinian evaporites from the Mediterranean and Red Seas. *Marine Geology*, **26**(1-2): 71-72.

Iaccarino, S.M., 1985. Mediterranean Miocene and Pliocene planktic foraminifera. In: Bollii, H.M., *et al.* (Ed.). *Plankton Stratigraphy,* Cambridge University Press: 283–314.

Iaccarino, S.M., Bossio, A., 1999. Paleoenvironment of uppermost Messinian sequences in the Western Mediterranean (Sites 974, 975, and 978). *Proceedings of the Ocean Drilling Program, Scientific Results*, **161**: 529-541.

Iaccarino, S.M., Castradori, D., Cita, M.B., Di Stefano, E., Gaboardi, S., McKenzie, J.A., Spezzaferri, S., Sprovieri, R., 1999. The Miocene-Pliocene boundary and the significance of the earliest Pliocene flooding in the Mediterranean. *Memorie della Societa Geologica Italiana*, **54**: 109-131.

Janofske, D., 1996. Ultrastructure types in recent "Calcispheres". *Bulletin de l'Institut océanographique*, **14**(4): 295-303.

Janofske, D., 2000. *Scrippsiella trochoidea* and *Scrippsiella regalis*, nov. comb. (Peridiniales, Dinophyceae): a comparison. *Journal of Phycology*, **36**: 178–189.

Janofske, D., Karwath, B., 2000. Oceanic calcareous dinoflagellates of the equatorial Atlantic Ocean: cyst-theca relationship, taxonomy and aspects on ecology. *Berichte FB Geowissenschaften*, Universität Bremen, **152**: 93–133.

Karwath, B., Janofske, D., Tietjen, F., Willems, H., 2000. Temperature effects on growth and cell size in the marine calcareous dinoflagellate Thoracosphaera heimii. *Marine Micropaleontology*. **39**: 43-51.

Keupp, H., Versteegh, G. 1989. Ein neues systematisches Konzept für kalkige Dinoflagellaten-Zysten der Subfamilie Orthopithonelloideae Keupp 1987. *Berliner Geowissenschaftliche Abhandlungen*, **106**(A): 207–219.

Keupp, H., Bellas, S.M., Frydas, D., Kohring, R., 1994. Aghia Irini, ein Neogenprofil auf der Halbinsel Gramvoússa/NW-Kreta. *Berliner Geowissenschaftliche Abhandlungen*, **13**(E): 469-481.

Keupp, H., Kohring, R., 1999. Kalkige Dinoflagellatenzysten aus dem Obermiozän (NN 11) W von Rethymnon (Kreta). *Berliner Geowissenschaftliche Abhandlungen*, **30**(E): 33-53.

Kohring, R., 1993(a). Kalkdinoflagellaten aus dem Mittel- und Obereozän von Jütland (Dänemark) und dem Pariser Becken (Frankreich) im Vergleich mit anderen Tertiär-Vorkommen. *Berliner Geowissenschaftliche Abhandlungen*, **6**(E): 1-164.

Kohring, R., 1993(b). Kalkdinoflagellaten-Zysten aus dem unteren Pliozän von E-Sizilien. *Berliner Geowissenschaftliche Abhandlungen*, **9**(E): 15-23.

Kohring, R., 1997. Calcareous dinoflagellate cysts from the Blue Clay formation (Serravalian, Late Miocene) of the Maltese Islands. *Neues Jahrbuch, Geologisch-Paläontologische Mitteilungen*, **3**: 151-164.

Kouwenhoven, T.J., Morigi, C., Negri, A., Giunta, S., Krijgsman, W., Rouchy, J.-M., 2006. Paleoenvironmental evolution of the eastern Mediterranean during the Messinian: constraints from integrated microfossil data of the Pissouri Basin (Cyprus). *Marine Micropaleontology*, **60**: (1): 17-44.

Krijgsman, W., Blanc-Valleron, M.-M., Flecker, R., Hilgen, F.J., Kouwenhoven, T.J., Merle, D., Orszag-Sperber, F., Rouchy, J.-M., 2002. The onset of the Messinian salinity crisis in the eastern Mediterranean (Pissouri Basin, Cyprus). *Earth Planetary Science Letters*, **194**: 299-310.

Logan, A., Bianchi, C.N. Morri, C., Zibrowius, H., 2004. The present-day Mediterranean brachiopod fauna: diversity, life habits, biogeography and paleobiogeography. *Scientia Marina*, **68**(1): 163-170.

Loget, N., Van Den Driessche, J., 2006. On the origin of the Strait of Gibraltar. *Sedimentary Geology*, **188-189**: 341-356.

McKenzie, J.A., Sprovieri, R., 1990. Paleoceanographic conditions following the earliest Pliocene flooding of the Tyrrhenian Sea. *Proceedings of the Ocean Drilling Program, Scientific Results*, **107**: 405-414.

Meier, S.K.J., Willems, H., 2003. Calcareous dinoflagellate cysts from surface sediments of the Mediterranean Sea: distribution patterns and influence of main environmental gradients. *Marine Micropaleontology*, **48**: 321-354.

Meier, K.J.S., Zonneveld, K.A.F., Kasten, S., Willems, H., 2004. Different nutrient sources forcing increased productivity during eastern Mediterranean S1 sapropel formation as reflected by calcareous dinoflagellate cysts. *Paleoceanography*, **19**: 1-12.

Meier, K.J.S., Young, J., Kirsch, M., Feist-Burkhardt, S., 2007. Evolution of different life-cycle strategies in oceanic calcareous dinoflagellates. *European Journal of Phycology*, **42**(1): 81-89.

Meijer, P.T., Krijgsman, W., 2005. A quantitative analysis of the desiccation and re-filling of the Mediterranean during the Messinian Salinity Crisis. *Earth and Planetary Science Letters*, **240**: 510-520.

Meijer, P.T., Tuenter, E., 2007. The effect of precession-induced changes in the Mediterranean freshwater budget on circulation at shallow and intermediate depth. *Journal of Marine Systems*, **68**: 349-365.

Orszag-Sperber, F., 2006. Changing perspectives in the concept of ''Lago-Mare'' in Mediterranean Late Miocene evolution. *Sedimentary Geology*, **188/189**: 259–277.

Orszag-Sperber, F., Rouchy, J.-M., 1979. Le Miocène terminal et le Pliocène inférieur au Sud de Chypre. *Livret Guide, 5ème Séminaire sur le Messinien*, Chypre, PIGC: 117.

Orszag-Sperber, F, Rouchy, J.-M., Blanc-Valleron, M.-M., 2000. La transition Messinien–Pliocène en Méditerranée orientale (Chypre): la période du Lago-Mare et sa signification. *Comptes Rendus de l'Académie des Sciences de Paris*, **331**: 490-493.

Pierre, C., Rouchy, J.-M., Blanc-Valleron, M.-M., 1998. Sedimentological and stable isotope changes at the Messinian/Pliocene boundary in the Eastern Mediterranean (Holes 968A, 969A, and 969B). *Proceedings of the Ocean Drilling Program, Scientific Results*, **160**: 3-8.

Pierre, C., Caruso, A., Blanc-Valleron, M.-M., Rouchy, J.-M., Orzsag-Sperber, F., 2006. Reconstruction of the paleoenvironmental changes around the Miocene–Pliocene boundary along a West–East transect across the Mediterranean. *Sedimentary Geology*, **188-189**: 319-340.

Richter, D., Vink, A., Zonneveld, K.A.F., Kuhlmann, H., Willems, H., 2007. Calcareous dinoflagellate cyst distributions in surface sediments from upwelling areas off NW Africa, and their relationships with environmental parameters of the upper water column. *Marine Micropaleontology*, **63** (3-4): 201-228.

Rohling, E.J., Hilgen, F.J., 1991. The eastern Mediterranean climate at times of sapropel formation: a review. *Geologie en Mijnbouw*, **70**: 253-264.

Rossignol-Strick, M., 1985. Mediterranean Quaternary sapropels, an immediate response of the African monsoon to variation of insolation. *Palaeogeography, Palaeoclimatology., Palaeoecology*, **49**: 237-263.

Rossignol-Strick, M., 1987. Rainy periods and bottom water stagnation initiating brine accumulation and metal concentrations: 1. the Late Quaternary. *Paleoceanography*, **2**: 333-360.

Rouchy, J.-M., Pierre, C., 1979. Données sédimentologiques et isotopiques sur les gypses des séries évaporitiques messiniennes d'Espagne méridionale et de Chypre. *Revue de Geologie Dynamique et de Geographie Physique*., **21**(4): 267–280.

Rouchy, J.-M., Orszag-Sperber, F., Blanc-Valleron, M.-M., Pierre, C., Rivière, M., Combourieu-Nebout, N., Panayides, I., 2001. Paleoenvironmental Changes at the Messinian–Pliocene Boundary in the Eastern Mediterranean (Southern Cyprus basins): Significance of the Messinian Lago–Mare. *Sedimentological Geology*, **145**: 93–117.

Rouchy, J.-M., Caruso, A., Pierre, C., Blanc-Valleron, M.-M., Bassetti, M.A., 2007. The end of the Messinian salinity crisis: evidences from the Chelif Basin (Algeria). *Palaeogeography, Palaeoclimatology, Palaeoecology*, **254**(3-4): 386-417.

Ruddiman, W.F., Raymo, M.E., Martinson, D.G., Clement, B.M., Backman, J., 1989. Pleistocene evolution: northern hemisphere ice sheets and North Atlantic Ocean. *Paleoceanography* **4**: 353–412.

Seidenkrantz, M.S., Kouwenhoven, T.J., Jorissen, F.J., Shackleton, N.J., van der Zwaan, G.J., 2000. Benthic foraminifera as indicators of changing Mediterranean–Atlantic water exchange in the late Miocene. *Marine Geology*, **163**: 387-407.

Spezzaferri, S., Cita, M.B., McKenzie, J.A., 1998. The Miocene/Pliocene Boundary in the eastern Mediterranean: Results from sites 967 and 9691. *Proceedings of the Ocean Drilling Program, Scientific Results*, **160**:9-28.

Steenbrink, J., Hilgen, F.J., Krijgsman, W., Wijbrans, J.R., Meulenkamp, J.E., 2006. Late Miocene to Early Pliocene depositional history of the intramontane Florina-Ptolemais-Servia Basin, NW Greece: Interplay between orbital forcing and tectonics. *Palaeogeography, Palaeoclimatology, Palaeoecology*, **238**: 151-178.

Stow, D.A.V., Braakenburg, N.E., Xenophontos, C., 1995. The Pissouri Basin fan-delta complex, southwestern Cyprus. *Sedimentary Geology*, **98**: 254-262.

Streng, M., Hildebrand-Habel, T., Willems, H., 2002. Revision of the genera Sphaerodinella Keupp and Versteegh, 1989 and Orthopithonella Keupp in Keupp and Mutterlose, 1984 (Calciodinelloideae, calcareous dinoflagellate cysts). *Journal of Paleontology*, **76**(3): 397-407.

Streng, M, Hildebrand-Habel, T., Willems, H., 2004. Long-term evolution of calcareous dinoflagellate associations since the Late Cretaceous: comparison of a high- and a low-latitude core from the Indian Ocean. *Journal of Nannoplankton Research*, **26**: 13-45.

Tanimura, Y., Shimada, C., 2004. Calcareous dinoflagellates from a northwestern Pacific sediment trap and their paleoceanographic implications. *Micropaleontology*, **50**: 343-356.

Van der Laan, E., Snel, E., de Kaenel, E., Hilgen, F.J., Krijgsman, W., 2006. No major deglaciation across the Miocene-Pliocene boundary: Integrated stratigraphy and astronomical tuning of the Loulja sections (Bou Regreg area, NW Morocco). *Paleoceanography*, **21**: PA3011, doi:10.1029/2005PA001193.

Van Harten, D., 1984. A model of estuarine circulation in the Pliocene Mediterranean based on new ostracod evidence. *Nature*, **312**: 359 – 361.

Vink, A., Zonneveld, K.A.F., Willems, H., 2000. Distributions of calcareous dinoflagellate cysts in surface sediments of the western equatorial Atlantic Ocean, and their potential use in Palaeoceanography. *Marine Micropaleontology*, **38**: 149-180.

Vink, A., Brune, A., Holl, C., Zonneveld, K.A.F., Willems, H., 2002. On the response of calcareous dinoflagellates to oligotrophy and stratification of the upper water column in the equatorial Atlantic Ocean. *Palaeogeography, Palaeoclimatology, Palaeoecology*, **178**(1-2): 53-66.

Vink, A., 2004. Calcareous dinoflagellate cysts in South and equatorial Atlantic surface sediments: diversity, distribution, ecology and potential for palaeoenvironmental reconstruction. *Marine Micropaleontology*, **50**: 43-88.

Wendler, I., Zonneveld, K.A.F., Willems, H., 2002a. Calcareous cyst-producing dinoflagellates: ecology and aspects of cyst preservation in a highly productive oceanic region. In: Clift, P.D., Kroon, D., Geadicke, C., Craig, J. (Editors). The tectonic and climatic evolution of the Arabian Sea region. *Geological Society Special Publication*, **195:** 317-340.

Wendler, I., Zonneveld, K.A.F., Willems, H., 2002b. Oxygen availability effects on early diagenetic calcite dissolution in the Arabian Sea as inferred from calcareous dinoflagellate cysts. *Global and Planetary Change*, **34**: 219-239.

Wendler, I., Zonneveld, K.A.F., Willems, H., 2002c. Production of calcareous dinoflagellate cysts in response to monsoon forcing off Somalia: a sediment trap study. *Marine Micropaleontology*, **46**: 1-11.

Willett, S.D., Schlunegger, F., Picotti, V., 2006. Messinian climate change and erosional destruction of the central European Alps. *Geology*, **34**(8): 613–616.

Zonneveld, K.A.F., Brune, A., Willems, H., 2000. Spatial distribution of calcareous dinoflagellates in surface sediments of the South Atlantic Ocean between 13°N and 36°S. *Review of Palaeobotany and Palynology*, **111**: 197-223.

Zonneveld, K.A.F., 2004. Potential use of stable oxygen isotope composition of Thoracosphaera heimii for upper water column (thermocline) temperature reconstruction. *Marine Micropaleontology*, **50**(3/4): 307-317.

Zonneveld, K.A.F., Meier, K.J.S., Esper, O., Siggelkow, D., Wendler, I., Willems, H. 2005. The (palaeo-) environmental significance of modern calcareous dinoflagellate cysts: a review. *Paläontologische Zeitschrift*, **79**(1): 61-77.

Chapter 4

Calcareous dinoflagellate cyst distribution and their environmental implications preceding and following the Messinian Salinity Crisis in the Caltanissetta Basin, Sicily

***Katarzyna-Maria Bison[1*] & Helmut Willems[1]**

[1]*Division of Palaeontology, University of Bremen, FB 5, Geowissenschaften, Postfach 330440, D-28334 Bremen, Germany*

*Corresponding author: kbison@uni-bremen.de

Palaeogeography, Palaeoclimatology, Palaeoecology
(to be submitted to)

Abstract

During the Messinian Salinity Crisis (MSC), at the end of the Miocene, large parts of the Mediterranean basin dried up or developed into saline evaporation basins. This mainly was due to the restriction of the Mediterranean/Atlantic gateway. The hydrological and climatic processes leading to and directly following the MSC are still a matter of debate. To shed more light on these periods we reconstructed the environmental evolution by means of the calcareous dinoflagellate cyst assemblages deposited in the Caltanissetta Basin prior to (~7.53-~5.97 Ma) and following the MSC (~5.33-~5.2 Ma). This study is part of a larger project looking at environmental gradients and their evolution, through time and space, on an east-west transect through the Mediterranean using calcareous dinoflagellate cysts as a proxy. Here we present the results of our study from the Caltanissetta Basin (Sicily) which followed our study from the Pissouri Basin (Cyprus). For the investigated stratigraphic interval we identified seven major phases: Prior to the MSC, during the upper Tortonian (~7.53 - ~7.51 Ma), we reconstructed strongly oligotrophic conditions modified by shifts in temperature and salinity with a distinct temperature drop at the top (~7.51 Ma). Representative main species are *Calciodinellum albatrosianum*, *Pernambugia tuberosa* and *Lebessphaera urania.* During the lower Messinian dominance of *C. albatrosianum* indicates rather stable warm and oligotrophic surface waters, interrupted by a cooling event at 7.17 Ma. During the upper Messinian (~6.78 - ~5.97 Ma) strong assemblage changes indicate alternating humid (*L. granifera* dominance) and warm arid (*C. albatrosianum* dominance) conditions, intermittent by short cooling phases with enhanced inflow of Atlantic water (dominance of *T. heimii*). Samples barren or with poor assemblages, indicate progressive restriction of the basin well before the MSC. The major change in the dinoflagellate cyst community, took place with the onset of the Pliocene (5.33 Ma) just after the reestablishment of fully marine conditions, shown by the significant replacement of *C. albatrosianum* through *L. granifera*. This early Pliocene change in dominance indicates eutrophication of the photic zone which is most likely triggered by intensified continental runoff in combination with local upwelling, the latter activated by an early estuarine circulation system during a cooler and more humid climate phase. This short-lasting early Pliocene event is immediately followed by an episode of ~120 ka with varying sea surface temperatures and nutrient levels concluding with the recovery of *C. albatrosianum* and the replacement of *L. granifera*. This final change in the dinoflagellate cyst record probably indicates a further reversal of the water circulation towards an anti-estuarine system. This terminal change we relate to a drier and somewhat warmer climate period, finally leading to a notable nutrient loss in the surface water. At the same time, a shallowing of the basin is reflected in a slight increase of neritic species, perhaps as a response to regional uplift. Overall, the early Pliocene evolution of the calcareous dinoflagellate cysts reflects the reorganization of the Mediterranean basin in four more or less distinct phases. The terminal Pliocene dinoflagellate cyst community still differs from those of the Mediterranean today. Some of the most common species of the present Mediterranean Sea (*Thoracosphaera heimii*, *Lebessphaera urania*, *Calciodinellum levantinum*) are still strongly underrepresented or even missing *(Calciodinellum elongatum*). On the other hand *C. albatrosianum* – the dominating species of our final Pliocene record - today accounts not more than 5% of the Mediterranean dinoflagellate association. When exactly the full establishment of the modern calcareous dinoflagellate association took place remains an issue for future studies. We can only say that this change must have taken place after the earliest Pliocene time. The main environmental trends as reflected by our dinoflagellate cyst record are in accordance with the observations of our first study from the eastern Mediterranean Pissouri Basin (Cyprus).

Keywords: Messinian Salinity Crisis, early Pliocene, Central Mediterranean, Caltanissetta Basin, Palaeoenvironment, Calcareous dinoflagellates

Introduction

The detailed history of the MSC is still poorly understood, such as the complex interference between the different environmental variables and processes, for example the balance between saltwater inputs from the Atlantic, freshwater input from rivers and evaporation/precipitation, as well as regional tectonic activity and eustasy. The tectonic processes in the Gibraltar strait region are mostly responsible for the restriction of the Atlantic water inflow and thus caused the Mediterranean-wide sea level drop, which eventually resulted in the Messinian Salinity Crisis. The interference of the climatic, oceanographic and tectonic processes and its environmental consequences are up to now only poorly understood, both temporally and spatially.

The analysis of the calcareous dinoflagellate cyst distributions can help to gain more insight into these issues. These organisms represent a group of phytoplankton which lives in the photic zone of the world's oceans. They react sensitive to environmental conditions such as sea surface temperature (SST), salinity, nutrient availability and stratification (e.g. Wendler *et al.* 2002a, b, c; Vink, 2004; Zonneveld, 2004; Zonneveld et al., 2005; Meier and Willems, 2003; Richter et al., 2007). Accordingly, shifts in fossil calcareous dinoflagellate cyst assemblages reflect modifications of past conditions in the upper water column, and thus provide direct information on palaeoceanographic and climatic changes. Therefore their response to palaeoclimatic and palaeoenvironmental changes and their good fossil preservation make them a useful proxy for the palaeoenvironmental reconstruction of the MSC. So far only few studies used this approach (e.g. Keupp and Kohring, 1993; Kohring, 1993, 1997; Meier et al. 2004). Bison et al. (2007; 2009) used this approach to unravel the environmental evolution of the MSC in the Pissouri Basin on Cyprus. We now moved westward to a time-equivalent sediment succession from the Caltanissetta Basin on Sicily. Comparison of the successions from Cyprus and Sicily may help to separate local from regional environmental changes.

Project Objectives

The main objective of our study is to reconstruct the changes of the environmental conditions in the time preceding and following the MSC, based on the distribution pattern of calcareous dinoflagellate cysts. In addition we want to increase our understanding of the evolution of calcareous dinoflagellates, spatially and temporally. For this we took an inventory of the dinoflagellate cyst association prior to and after the MSC. The extent of the Messinian crisis on the development of the Neogene Mediterranean calcareous dinoflagellate cyst association in the Mediterranean realm will be addressed by the following questions.

- How did the calcareous dinoflagellate cyst association react to this drastic event and the accompanying environmental changes?
- What happened to the calcareous dinoflagellate association inhabiting the Mediterranean prior to the salinity crisis?
- Did the former Tethyan calcareous dinoflagellate association survive the crisis?
- How did this group respond to the re-establishment of normal marine conditions in the Pliocene?
- When did the modern Mediterranean dinoflagellate association establish?

By answering these questions we want to gain more insight into both, the history of the MSC and the evolution of this taxonomic group, spatially and temporally, and thus encouraging their position as a proxy for palaeoenvironmental reconstructions.

In the considered time interval the Mediterranean basin underwent substantial changes related to the MSC. These changes were mainly tectonically driven but superimposed by long- and short-term climatic changes as a response to the Milankovitch Cycles. Witnesses of the periodic short-term changes are the alternating marl and sapropel layers. The alternation is considered to be precession driven and reflects the succession of more humid (sapropels) and dryer periods (marls). These cycles are superimposed by large-scale climatic changes (Daux et al., 2006; Huesing et al., 2009). The orbitally induced climatic changes are modified by regional and supra-regional tectonic processes, all these effects being recorded in the sediments.

The analysis of marl and sapropel layers was included, whereas the intercalated diatomite layers are not involved in this study as they are suggested to contain no calcareous dinoflagellate cysts. Most studies have interpreted the diatomite layers as first discrete expressions of basin restriction and sea level fluctuation toward the MSC (e.g. McKenzie et al., 1979; Meulenkamp et al., 1979; Gersonde and Schrader, 1984; Thunell et al., 1984; Van der Zwaan and Gudjonsson, 1986; Grasso et al., 1991; Suc et al., 1995; Hüsing et al., 2009). Within our Sicily record (Falconara section) first diatomite deposits occur at about 6.47 Ma.

Study area

The Miocene/Pliocene interval studied here is composed of the Falconara, Gibliscemi and Eraclea Minoa sections, all deposited in the Neogene-Quaternary Caltanissetta Basin on Sicily (Fig. 1).

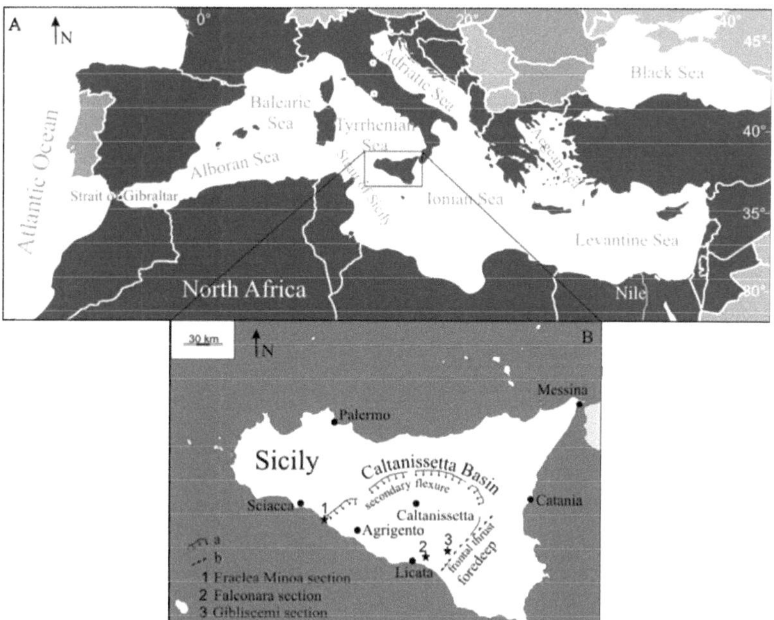

Fig. 1: **(A)** Generalised map of the Mediterranean Sea with nomenclature of the major sub-basins and straits. Sicily is located in the central Mediterranean Sea. **(B)** Location map of the three studied sections of the Caltanissetta Basin on Sicily: 1 Eraclea Minoa, 2 Falconara and 3 Gibliscemi (modified after Londeix et al. 2007). a = secondary flexure; b = modern thrust front (from Broquet et al., 1984; Pedley and Grasso, 1993).

Based on palaeobathymetric reconstructions of the pre-evaporitic sections Falconara and Gibliscemi, the Sicilian sediments have been deposited within a relatively deep basinal setting in the Caltanissetta Basin with estimated water depths of approximately 1200 m in the early Messinian (Kouwenhoven et al., 2003; Kouwenhoven and Van der Zwaan, 2006; Krijgsman and Meijer, 2008). The evaporitic succession of Sicily is therefore regarded as indicative for a relatively deep peripheral basin containing the best analogue MSC succession of the deep Mediterranean basins (Rouchy and Caruso, 2006), with moderate post-depositional tectonic deformation (Hilgen et al., 2000). The sections encompass the entire history of the environmental conditions before, during and after the MSC.

According to Londeix et al. (2007), prior to the crisis, normal marine conditions are reflected in the blue-grey marls (Terravecchia Formation) of the late Tortonian; the pre-evaporitic Messinian sediments (lower Tripoli Formation) consist of alternating layers of marine marls, clays, sapropels and laminated diatomites, reflecting highly variable conditions within an increasingly restricted basin; the strongly restricted marine evaporitic phase, bears the none-marine Lago Mare sequence (upper Tripoli Formation) at the top of the Messinian. This is followed by the Early Pliocene normal marine calcareous marls (Trubi formation).

Gibliscemi section

The Gibliscemi section is located in the southern part of Sicily (Italy), 25 km east of Falconara, on the southern slope of Monte Gibliscemi (Hilgen et al., 1995). The studied sequence represents the upper Serravallian up to the lower Messinian (Kouwenhoven et al., 2003). The sediments consist of open marine, cyclically arranged marly and sapropelic sequences, deposited within a relative deep basinal setting between about 800 and 1200m water depth (Hilgen et al., 2000; Kouwenhoven et al., 2003). Tectonic activity and low angle shear planes, particularly in the lower parts of the section, led to a stratigraphic reduction; as a consequence no continuous undisturbed succession exist (e.g. Hilgen et al., 2005).

Falconara section

The Falconara section is located on southern Sicily, 3.5 km NW of the castle Falconara, between Licata and Gela (Hilgen and Krijgsman, 1999). The section encompasses cyclically bedded layers (tripartite cycles) of the Tripoli formation, consisting of homogenous marls, laminated sapropels and diatomites (Hilgen and Krijgsman, 1999). It is overlain by the carbonate cycles (cycle 50 to 52) of the Calcare di Base formation (Sprovieri et al., 1996a), which represents the basal part of the so called Gessoso Solfifera Formation (Decima and Wezel, 1973), encompassing the evaporitic deposits and the Lago Mare facies. The Falconara section contains 45 cycles of the Tripoli formation which reach up to 2 m thickness (Sprovieri et al., 1996a, b; Sprovieri et al., 1997; Bellanca et al., 2001). In the lower segment some cycles are absent as a result of tectonic activity. A correlation of the Falconara and Gibliscemi section established a total of 49 cycles for the Tripoli formation (Hilgen and Krijgsman, 1999).

Eraclea Minoa section

The Eraclea Minoa section is located on the southwest coast of Sicily (near the community of Eraclea Minoa) (e.g. Van Couvering et al., 2000). The section provides one of the most complete sedimentary successions of the MSC including the lower evaporites, the upper evaporites, the transitional Lago Mare facies and the early Pliocene Trubi Formation, straddling the Miocene-Pliocene boundary. The boundary marks the return to fully marine conditions following the Messinian salinity crisis in the Mediterranean region (Cita and Gartner, 1973; Ruggieri and Sprovieri, 1976). The deposits of Eraclea Minoa form the base of the Rosella Composite Section and thus the base of the Pliocene (Langereis and Hilgen, 1991). They consist of rhythmically bedded Trubi marls. The Eraclea Minoa section also concerns as a stratotype section (GSSP) for the Messinian-Zanclean boundary, with an orbitally calibrated age of the Zanclean base of 5.33 Ma (Langereis and Hilgen, 1991). The contact of the well exposed basal Zanclean is pretty obvious and consists of white Trubi marls, which directly overly the dark brown sandy Arenazzolo deposits (Langereis and Hilgen, 1991).

Material

The 63 samples studied cover the stratigraphic interval from the upper Tortonian (~7.53—~7.51 Ma) through the pre-evaporitic Messinian (7.24—~5.97 Ma), up to the onset of the evaporate formation (Calcare di Base), ending with the first about 120 ka of the basal Pliocene (~5.33-~5.21 Ma). Sediments encompassing the evaporite and Lago Mare bearing deposits from 5.96-5.33 Ma (Hilgen et al. 2007), have not been studied since these environments do not contain calcareous dinoflagellates. The studied samples derive from successions of the Gibliscemi, Falconara and Eraclea Minoa section, all situated in the Caltanissetta Basin on Sicily. As parts of the Gibliscemi section are tectonically disturbed and cover various shear planes no continuous undisturbed sequence exists (Langereis and Hilgen, 1991; Hilgen et al. 2000). To obtain a relatively undisturbed succession, we therefore build a composite of the aforementioned sections.

Our samples from the Gibliscemi section comprise the uppermost Tortonian and lower Messinian interval (~7.53 to ~7.11 Ma), belonging to the Licata Formation of southern Sicily. The formation represents the distal equivalent of the Terravecchia Formation (Butler et al., 1999). From the Falconara section we obtain our upper Messinian samples which cover the time interval from ~7.04 to ~ 5.97 Ma, encompassing the Tripoli Formation and two samples from the Calcare di Base carbonates of the Gessoso Solfifera Formation. The early Pliocene first about 120 ka (~5 m) investigated here derive from the Trubi Formation of the Eraclea Minoa section.

Method

The laboratory procedure is according to Bison et al. (2007). Shortly, ~0.5 g of dried sediment was disaggregated, and the 20-65 µm fraction was isolated by sieving with tap water. An aliquot of this fraction was mounted on a scanning electron microscope (SEM) stub for further quantitative and qualitative analyses. Since light microscopic analysis allows assessment of the crystallographic orientation of the c-axes of the wall crystals it was used to verify the identity of taxa which are difficult to identify by means of SEM. For diversity calculations the Shannon-Weaver index (Shannon and Weaver, 1948) was used.

Results

Almost all samples contain mostly well-preserved calcareous dinoflagellate cysts, except two samples from the lower and three samples from the uppermost part of the Messinian which are barren (Fig. 2; Appendix 2). In addition, in some samples calcareous dinoflagellate cysts are extremely rare, with less than 10 cysts/mg sediment (Appendix 2). We distinguished seven stratigraphic intervals; three (I1, I2, I3) from the Tortonian/Messinian and four (I4, I5, I6, I7) from the Pliocene.

General trends

Species composition

C. albatrosianum generally dominates the Tortonian/Messinian assemblages with about 61% on average, while *L. granifera* is the main species (~48% on average) of the Pliocene dinoflagellate association, where *C. albatrosianum* reaches only 22% on average (Fig. 2). Together these species form almost 70% of the total dinoflagellate cyst community.

C. albatrosianum is continuously present throughout the stratigraphic interval investigated, except in those samples barren of calcareous dinoflagellate cysts (Fig. 2). In contrast, *L. granifera* occurs only sporadically and with low numbers (~9% on average) throughout the Tortonian/Messinian period, albeit with a slight upward increase. With the beginning of the Pliocene *L. granifera* significantly increases and becomes the dominant species of the dinoflagellate cyst community.

The common but scarce species *Calciodinellum levantinum*, *Calciodinellum operosum*, *Caracomia stella*, *Lebesphaera urania*, *Pirumella parva*, *Pirumella tuberosa*, *Thoracosphaera heimii* and *Rhabdothorax* spp. as well form an important part of the dinoflagellate community, even though with strongly varying relative and absolute abundances (Fig. 2). Hereof, *C. levantinum* is restricted to the Pliocene, whereas *P. tuberosa* is more or less limited to the Tortonian (I1) and lower Messinian (I2).

The remaining species (*Calciodinellum limbatum*, *Melodomuncula berlinensis*, *Calcigonellum infula*, *Ruegenia oranensis*, *Praecalcigonellum polymorphum*, *Pirumella rhombica*, *Praecalcigonellum schizosaeptum*, *Pirumella sicelis*, sp.1), occur only sporadically and with very low numbers (> 1 cysts/mg) throughout the studied interval, with most of them even restricted to the Pliocene (Fig. 2; Appendix 1). Almost all species of this rare group are missing during the upper Messinian (I3), except *M. berlinensis*, which peaks (14%) in the upper part (Fig. 2; Appendix 2).

<u>*Cyst concentrations*</u>

Cyst concentrations are highest (average ~43 cysts/mg) during the lower Messinian interval (I2). The upper Messinian (I3) assemblages are marked by strongly varying values. Here rather moderate average cyst concentrations (~26 cysts/mg) are intermittent by samples containing concentration highs (up to 290 cysts/mg) and lows (<1 cyst/mg) up to samples barren of calcareous dinoflagellate cysts. The upper Tortonian (I1) and the complete Pliocene interval both show moderate total cyst numbers (~12 cysts/mg) (Fig. 2).

<u>*Cyst concentrations with respect to the lithology*</u>

During the upper Messinian maximum cyst concentrations are usually positively related to sapropel layers, whereas during the Tortonian (I1) and lower Messinian highest cyst concentrations are generally correlated to marl layers. In the Pliocene sequence no clear abundance trend in respect to sapropel/marl deposits became apparent.

Diversity

Diversity strongly varies throughout the studied interval, as well as inside the distinguished phases (Fig. 2). Highest diversities were recorded from the upper Tortonian (average H – 2.15) and the upper Messinian interval (I3) (average H = 1.59). Hereof, the latter shows extreme and high frequent alternating diversity levels, varying between maxima of e.g. H = 4.61 and minima of e.g. H = -1.95 (Fig. 2; Appendix 2). Lowest diversities occur during the lower Messinian (I2) (average H = 0.21). In general, lower diversities are associated with increased cyst concentrations and vice versa, being most significant during the prominent abundance peaks of the upper Messinian interval (I3) (Fig. 2).

Species richness

On total, species richness varies between two and twelve species per sample. Highest species numbers were found in the Pliocene assemblages, which generally contain more than 8 species per sample, except the lowermost sample which only holds 5 species. Lowest species numbers occur during the upper Tortonian, with five species at most. Strongest fluctuations in species richness (1–8) characterise the upper Messinian (I3), whereas relative stable and moderate species records are typical for the lower Messinian (I2).

Characteristics of the selected intervals

Interval 1 (upper Tortonian; ~7.53–~7.51Ma)

C. albatrosianum - P. tuberosa - L. urania - association

The most important species of this interval are *C. albatrosianum* (average 62%), *P. tuberosa* (average 24%) and *L. urania* (average 8%). *C. albatrosianum* dominates the lower part (~73%), but significantly decreases in the topmost sample (33%). *L. urania* is more or less restricted to the first sample where it reaches a maximum of ~29%. In the subsequent part it is very scarce or absent. *P. tuberosa* has its maximum (~67%) in the topmost sample. Cyst concentrations are moderate to low (average 12 cysts/mg) with a significant drop at the top (0.9 cysts/mg). Diversity is moderate to high (average H = 2.15), but associated with a relative low species richness (2-5).

Interval 2 (lower Messinian; 7.24 Ma-6.81 Ma)

C. albatrosianum – association

The significant increase and dominance (average 83%) of *C. albatrosianum* distinguishes this sequence from the previous one. Also the notable rise in absolute cyst concentration (average 42 cysts/mg) primarily is caused by this species (Fig. 2). *P. tuberosa* distinctly decreases and is more or less missing throughout this interval (less than 1%). Conversely other species for the first time occur, such as *P. parva* (average 4%) and *C. stella* (average 1%). The previous one is the second most abundant species of this interval, together with *L. urania* (average 4%) (Fig. 2). Also worth to mention are two rather prominent abundance peaks of *P. parva* (at 7.24 Ma and 7.17 Ma), coinciding with lowest values of *C. albatrosianum* inside this interval (Fig. 2). Furthermore rather low diversities (average H = 0.19), associated with moderate species richness (3 – 8), typify this interval (Fig. 2; Appendix 2).

Interval 3 (upper Messinian; ~6.8 Ma–5.97 Ma)

C. albatrosianum – L. granifera – T. heimii - association

Strong fluctuations both, in absolute and relative cyst abundances, as well as in diversity characterise this interval (Fig. 2). It starts with a distinct drop in absolute cyst concentration (average ~26cysts/mg) followed by a succession containing samples barren of calcareous dinoflagellate cysts, up to those comprising striking abundance peaks which are caused mainly by two species, *C. albatrosianum* (e.g. 156 cysts/mg) and *T. heimii* (e.g. 291 cysts/mg). It concludes with a total loss of dinoflagellate cysts in the uppermost part (Fig. 2). *C. albatrosianum* on average (54%) still dominates the dinoflagellate association, although notably reduced. Besides, other species become more important. The species *L. granifera*, *T. heimii* and to a minor amount *Rhabdothorax* spp. temporary take over the supremacy (Fig. 2). The relative common lower Messinian species *L. urania* and *P. parva* are notably reduced and almost absent now particularly in the upper part of this interval. *P. tuberosa*, which already strongly declined in the lower Messinian interval is lacking now.

Interval 4 (basal Pliocene; 5.33 Ma; cycle 0)

L. granifera - association

Here, a prominent abundance peak of *L. granifera* (~31 cysts/mg) characterises the assemblage. The species significantly dominates (~91%) this earliest Pliocene dinoflagellate association. It is followed by minor fractions of *C. levantinum* (3%), *Rhabdothorax* spp. (3%) and *T. heimii* (~2%) (Fig. 2). *C. levantinum* for the first time occurs within the stratigraphic interval studied. The main difference of the early Pliocene dinoflagellate assemblage to that of the upper Tortonian/Messinian one is the considerable increase of *L. granifera* in both, absolute and relative abundance, simultaneously with a significant drop of *C. albatrosianum* (<1%). After its total drop during the uppermost Messinian cyst concentration increased again (34 cysts/mg) but diversity is rather low now (H = -0.78).

Interval 5 (Pliocene; ~5.33-5.31 Ma; cycle 1a, c, d, 2a)

L. granifera – C. stella – C. albatrosianum - association

This interval is initiated by a striking drop of *L. granifera* in both, absolute (3 cysts/mg) and relative (~51%) abundances, associated with a distinct decrease in average cyst concentration (~7 cysts/mg). *C. albatrosianum* again increases slightly (~18%) together with various other species, in particular *C. stella* (13%), *P. parva* (7%), *Rhabdothorax* spp. (7%), *L. urania* (5%) and *T. heimii* (5%). *C. levantinum* is present throughout the interval but with strongly varying abundances. Overall *L. granifera* still dominates (~38%) this interval which is now marked by a relative diverse (average H = ~1.9) and species-rich (9 - 12 taxa) dinoflagellate association.

Interval 6 (Pliocene; ~5.23-5.25 Ma; cycle 2c, 3a, c, 4a, a2, c)

L. granifera – P. parva - C. albatrosianum – T. heimii - association

Here, *L. granifera* increases again both, in absolute (average 7 cysts/mg) and relative (average 53%) abundances. At the same time *C. albatrosianum* distinctly decreases (average 9%) concurrently with *C. stella* (average 0.6%). *P. parva* (average 16%) notably increases accompanied by a minor increase of *T. heimii* (average 9%). Cyst concentration on average (~14 cysts/mg) notably increases once more coupled with a decrease in diversity (average H = ~1.1). In the middle part of this sequence a synchronous rise of various species – such as *L. granifera*, *C. albatrosianum*, *L. urania*, *P. parva* and *Rhabdothorax* spp. – cause an outstanding abundance peak (24 cysts/mg). Interestingly *T. heimii* significantly drops during this peak from a maximum of ~21% in the previous sample down to less than 1% (Fig. 2).

Interval 7 (Pliocene; ~5.2 Ma; cycle 6)

C. albatrosianum – P. parva – Rhabdothorax spp. – L. granifera association

With this topmost sample the dinoflagellate cyst community considerably changes again. The most remarkable shift is the drastic reduction of *L. granifera* in both, absolute (0.7 cysts/mg) and relative (8%) abundance, associated with a striking drop of the total cyst numbers (~8 cysts/mg). At the same time *C. albatrosianum* substantially increases (~59%) and takes over the supremacy again. Concurrently nearly all other main species notably decline, such as *C. stella*, *L. urania*, *P. parva*, *T. heimii* and *Rhabdothorax* spp. In contrast, some species of the rare group (*C. operosum*, *P. polymorphum* and *P. schizosaeptum*) slightly increase as well as *C. levantinum*. *C. stella* remains low but constant in absolute cyst numbers.

The most common taxa observed in the current study are documented in Fig. 2. Calcareous dinoflagellate counts of the total assemblages are given in the Appendix 2; for SEM pictures of the main species see Plate 1 and 2.

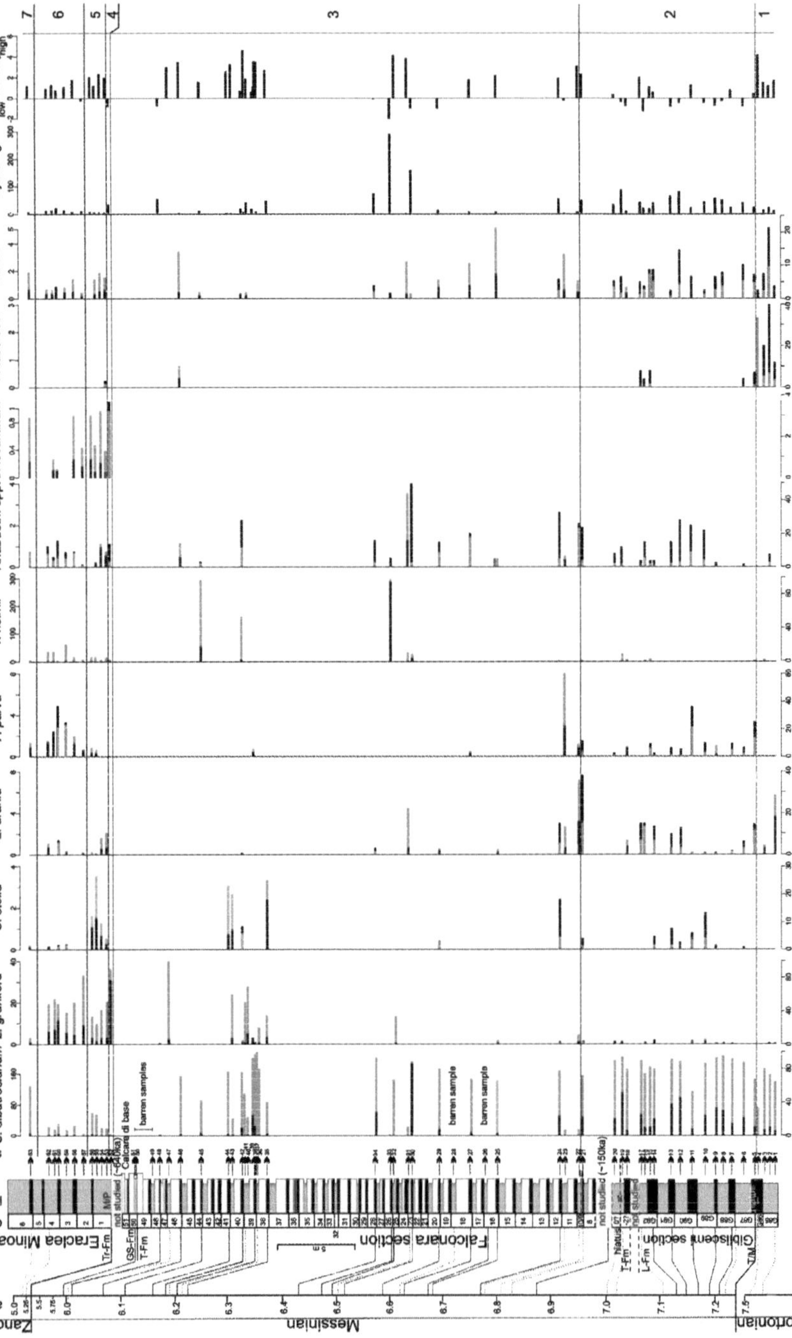

Fig. 2.

Stratigraphic composite profile of the uppermost Tortonian, pre-evaporitic Messinian and lower Pliocene (Zanclean; MPL1) succession of the three sections, Eraclea Minoa, Falconara and Gibliscemi, located in the Caltanissetta Basin on Sicily (modified after Krijgsman et al., 1995; Hilgen et al., 1995; 1999 Hilgen and Krijgsman, 1999 and Hilgen et al., 2007) and position of the samples (1 - 63) studied and the distinguished intervals (I1 - I7) are shown. Original sample and field numbers are listed in the Appendix 3, Table 1. Age, diversity, cysts abundances in cysts per mg sediment, relative abundance of the main cyst species (*Calciodinellum albatrosianum, Leonella granifera, Caracomia stella, Lebesphaera urania, Pirumella parva, T. heimii, Rhabdothorax* spp., *Calciodinellum levantinum, P. tuberosa*) and other cysts (*Calciodinellum limbatum, Calciodinellum operosum, Calcigonellum infula, Melodomuncula berlinensis, Praecalcigonellum schizosaeptum, Ruegenia oranensis, Pirumella* sp., *Pirumella sicelis*, undefined specimen), analysed in the samples, are displayed as percentage of the total cyst number against time.
The stratigraphic position in meter, related ages and corresponding precession cycles, the field numbers and original sample numbers are summarized in the Appendix 2, Table 1. T/M = Tortonian-Messinian boundary; M/P = Miocene-Pliocene boundary; L-Fm = Licata Formation; T-Fm = Tripoli Formation; GS-Fm = Gesosso Solfifera Formation; Tr-Fm = Trubi Formation.

Discussion

Environmental and ecological considerations on calcareous dinoflagellate taxa are based on both, modern data (e.g. Zonneveld et al., 2000; Karwath et al. 2000; Wendler et al., 2002a, b, c; Vink et al., 2002, Vink; 2004; Meier and Willems, 2003; Richter et al. 2007) and on palaeoecological/environmental assumptions (e.g. Keupp and Kohring, 1993, 1999; Keupp et al., 1994; Kohring, 1993a, b, 1997; Hildebrand-Habel and Willems, 2000; Meier et al., 2004; Bison et al., 2007, 2009). Most of the dinoflagellate species identified within this study are extant, providing valuable information on their ecology and environmental relations.

The dinoflagellate cyst assemblages in this study represent different environmental settings according to the environmental preferences of the main species. Consequently, the dinoflagellate cyst assemblages can be used to indicate specific surface water conditions. The main environmental parameters acting as limiting factors are sea surface temperature (SST), nutrient level, salinity, hydrodynamics and nearshore/offshore gradients. Therefore the observed variations of the dinoflagellate cyst communities through time reflect environmental changes in the Caltanissetta Basin both, as a response to the tectonically induced restriction in the Gibraltar area (MSC) and to short and long-term climatic changes in the Mediterranean realm. The environmental interpretation of the dinoflagellate cyst record here however becomes rather delicate due to the complex interaction of tectonic activity, eustasy and climatic change that all influence the sedimentation process in the Caltanissetta Basin during Late Miocene/Early Pliocene time (Londeix et al., 2007).

Species-environment relationships

The relationship between environmental factors and the distribution pattern of the main calcareous dinoflagellate species are described below. Taxa not involved in detail for the environmental interpretation are either very rare or have an uncertain or unknown environmental response.

<u>*C. albatrosianum: warm – oligotrophic - offshore*</u>

Today, the highest frequencies of *C. albatrosianum* occur in warm tropical and subtropical stable open oceanic environments characterised by low nutrient levels and low seasonal temperature gradients (e.g. Janofske and Karwath, 2000; Esper et al., 2000; Vink, 2004; Richter et al., 2007). Within warmer surface waters this species tolerates a wide range of nutrient and energetic levels and also occurs in coastal and neritic environments, but always with lower abundances than in the offshore regions (e.g. Wendler et al., 2002a, b; Vink, 2004; Richter et al., 2007). Hence *C. albatrosianum* is clearly not restricted to oligotrophic environments but obviously excellently adapted to it. A clear positive relationship with temperature and a reverse relationship with primary productivity in the surface water has been observed in several oceanographic studies (e.g. Esper et al., 2000; Wendler et al., 2002a; Vink, 2004; Richter et al., 2007). This is especially evident in high productive coastal regions influenced by upwelling and continental runoff. Here this species shows a clear offshore trend with an increase in abundance (e.g. Wendler et al., 2002a; Richter et al., 2007).

Temperature and low seasonality in SST are consequently the most important environmental factors affecting the distribution of this species (e.g. Esper et al., 2000; Vink, 2004; Richter et al., 2007). The scarcity of *C. albatrosianum* in the present Mediterranean Sea, which is characterised by strong temperature seasonality, supports this assumption. Here this species only reaches up to 5% of the dinoflagellate association (Meier and Willems, 2003). In addition, a positive correlation between increased frequencies of *C. albatrosianum* and the depth around the upper thermocline has been observed in the equatorial Atlantic (Janofske and Karwath, 2000), suggesting a preference for the environmental conditions at this depth.

However, within this work we use increased abundances and dominance of *C. albatrosianum* primarily as an indicator for stable warm and oligotrophic surface waters. Such conditions are usually associated with dry and warm climate periods. Consequently, lower percentages of this species we correlate with a general or seasonal decrease in sea surface temperature associated to more humid and cooler climate phases.

C. operosum: warm - oligotrophic - nearshore - offshore

The rare species, *C. operosum*, generally shows a similar geographic distribution pattern as *C. albatrosianum* (Vink, 2004; Richter et al., 2007) and thus might be linked to similar environmental characteristics. However, a more shelfward distribution of *C. operosum* has been observed at the coast off NW Africa by Richter et al., (2007). A restriction of this species to more coastal regions with a tendency to shallower environments was also derived from Mediterranean surface samples (Meier and Willems, 2003).

L. granifera: euryhaline - cool to warm – eutrophic – coastal

Today, increased numbers of *L. granifera* usually occur in high productive coastal regions of the tropical to subtropical Atlantic Ocean and in the Arabian Sea (e. g. Wendler et al., 2002a, b, c; Siggelkow et al., 2002; Vink, 2004; Richter et al., 2007). In the present highly oligotrophic present Mediterranean Sea *L. granifera* is very scarce and was so throughout the Holocene, except during sapropel S1 formation (Zonneveld et al., 2001; Meier and Willems, 2003). The sapropel S1 is considered to reflect enhanced productivity in the surface waters linked to increased freshwater runoff into the Mediterranean basin (e.g. Rossignol-Strick et al., 1982; Hilgen, 1991; De Lange et al., 1999; Rohling and Hilgen, 1991; Rohling, 1994; Meyers, 2006).

At present a clear positive relationship of elevated *L. granifera* cyst concentrations was observed for several regions where river runoff is substantial, such as at the Amazon and Orinoco outflow areas (Vink et al., 2001, 2002), the Gulf of Guinea which is strongly influenced by the Congo River (Vink, 2004) and the Indus Fan delta of the NE Arabian Sea (Wendler et al., 2002a). In the latter, in addition to the nutrient enrichment via the Indus River, deep water mixing brings large amounts of nutrients into the photic zone (Wendler et al., 2002a). Here, the exceptionally high abundances of *L. granifera* can be explained by eutrophication of the surface water through a combination of two processes, river runoff and deep water mixing. Actually, a clear positive relationship of extreme *L. granifera* peaks to coastal upwelling is documented from surface sediment and sediment trap studies off the north west African coast (Richter et al., 2007; Richter, 2009) and the south equatorial Atlantic (Vink, 2004). However, it may not be the upwelling process itself which stimulates *L. granifera* productivity but the accompanying environmental conditions, particularly increased nutrient/mineral availability.

Palaeoenvironmental studies of the upper Neogene eastern and central Mediterranean basin (Bison et al, 2007, 2009; this work) confirm these observations. Here, considerable peak occurrences of *L. granifera* at the beginning of the Pliocene epoch are strongly related to nutrient enrichment through a combination of both upwelling and increased continental runoff. As previously mentioned, an apparent positive correlation of elevated *L. granifera* productivity and nutrient enrichment through continental runoff is also shown by Meier et al. (2004) during the time of sapropel S1 formation. Here, cyst concentrations of *L. granifera* significantly increased during this time, but notably dropped before a maximum barium excess was reached in the sapropel layer (Meier et al., 2004). Meier et al. (2004) suggest a general decrease in the Nile River discharge, which is suggested to be the main nutrient/mineral source during sapropel S1 formation. In addition, we interpret this drop, before the end of sapropel formation, to reflect a modification in the nutrient quality linked to a change in or of the drainage area, emphasizing a clear response of *L. granifera* to specific nutrients or minerals.

One element which can be considered in this respect is iron, as it is an essential and limiting nutrient for phytoplankton growth, controlling overall biological productivity in marine environments (e.g. Martin et al., 1993; 1994; Morel et al., 1991; Geider and La Roche, 1994; Geider, 1999; Johnson et al., 1999; Jickells et al., 2005; Sarthou et al., 2003; Sundareshwar et al., 2003, 2007). The key sources of iron are wind transported dust, continental runoff and upwelling. However, as iron must be dissolved to be accessible to phytoplankton the most bioavailable iron is that supplied by wet transportation mechanisms, such as through river runoff, upwelling and wet-deposited dust (e.g. Lam and Bishop, 2008; Shi et al., 2009). Dry deposited dust on the other hand is very unreactive and not accessible to phytoplankton (Lam and Bishop, 2008; Shi et al., 2009) and first has to be converted into a bioavailable form. Furthermore, the consumption of iron is species specific and depends, for example, on individual growth rates (e.g. Geider and La Roche, 1994 and citations herein). Fast growing taxa like *L. granifera* for instance, probably have a higher iron exchange and thus may directly respond to iron availability in the surface water. In contrast, open oceanic species like *C. albatrosianum* are adapted to lower nutrient levels and likely also to lower trace mineral (iron) levels. Coastal and neritic species on the other hand generally have a higher nutrient/mineral uptake (see Brand et al., 1983).

Apart from nutrient depletion, the near absence of *L. granifera* in the modern Mediterranean Sea can be explained by relatively high salinity and/or strong seasonality in SST. Since *L. granifera* is well adapted to environments of enhanced river runoff, where salinity fluctuation is common, it obviously can tolerate a wide range of salinity levels, suggesting a euryhaline character. A positive correlation of this species with lowered salinities has been documented in several studies (Meier et al., 2004; Vink et al., 2000; Vink, 2004; Zonneveld et al., 2005). In addition, a preference of *L. granifera* for the upper photic zone between 10 and 50m water depth was documented from equatorial Atlantic surface water samples by Janofske and Karwath (2000). Tolerance to a wide temperature range can be inferred from its occurrence in both relatively cool surface waters of the upwelling regions and the warm equatorial and South Atlantic Oceans (Vink, 2004).

Altogether we think that the main environmental factor for increased cyst productivity of *L. granifera* is high and consistent nutrient and/or mineral availability in the surface water. This can be due to continental runoff, dust input, upwelling or ideally a combination of all of them. Furthermore we suggest coincident with Vink (2004) that *L. granifera* requires crucial nutrients and/or minerals, such as iron. Except for nutrient availability and high salinities we attribute *L. granifera* to a rather broad tolerance to varying environmental factors. Hence, in the following we use *L. granifera* as an indicator for surface water eutrophication, induced by continental runoff, upwelling or both.

T. heimii: Cool - eutrophic - DCM

Today, *T. heimii* is geographically widespread and overwhelmingly dominates the calcareous cyst producing dinoflagellate associations of the world's ocean, making up generally more than 90% of the dinoflagellate community (e.g. Höll et al., 1998; Janofske and Karwath, 2000; Karwath et al., 2000a; Wendler et al., 2002a, b, c; Tanimura and Shimada, 2004; Vink, 2004). It is found in various marine environments, from neritic to coastal and oceanic ones, displaying a broad tolerance to a wide range of environmental conditions (Vink et al., 2002; Vink, 2004). Besides broad environmental tolerance, high cyst reproduction rates (Inouye and Pienar, 1982; Tangen et al., 1982) are responsible for *T. heimii*'s worldwide dominance within the calcareous dinoflagellate community today. Highest abundances of *T. heimii* generally occur in somewhat cooler regions with temperatures ranging from 14°C to ~27°C (Zonneveld et al., 2000; Karwath et al., 2000a; Vink, 2004). A preference to rather cool water temperatures is also supported by growth experiments on *T. heimii* showing maximal cyst production rates at relatively low temperatures of about 16°C (Karwath et al., 2000b).

In general *T. heimii* prefers to live in relatively stable, well stratified waters within the lower photic zone at depth of about 50 to 100m which corresponds with the nutrient enriched deep chlorophyll maximum (DCM) layer (Karwath, 2000; Karwath et al., 2000a; Vink et al., 2002; Vink, 2004). In this layer - just below the thermocline - temperatures and light irradiance are relatively low but availability of nutrients is often greater than in the surface layer. Therefore, *T. heimii* prefers nutrient availability over light irradiance. Although *T. heimii* generally prefers well stratified waters, density stratification by river runoff appears to have an adverse effect on its productivity (Vink, 2004).

Within this work we link increased cyst concentrations and dominance of *T. heimii* to relatively cool and stratified waters associated with low nutrient availability in the upper surface layer but high nutrient availability at deeper parts (DCM layer) of the photic zone. *T. heimii*'s preference for depths with lower light irradiance differentiates it from *L. granifera*.

In contrast to *T. heimii's* present day overall dominance within the calcareous dinoflagellate community, in our Miocene/Pliocene interval it is remarkably underrepresented. The exact timing of its dominance in the world's oceans is still unclear. The first questionable and scarce occurrences of *T. heimii* can be traced back to the Upper Cretaceous (Hildebrand-Habel et al., 1999). In the eastern Mediterranean basin, the first sporadically increased occurrences of *T. heimii* have been described from the early Pliocene (Bison et al., 2007, 2009). In the central Mediterranean (Caltanissetta Basin, this work) first peak occurrences of *T. heimii* were recorded for the upper Messinian. It is still an open question why its advance to dominance required so much time, namely at least about 60 Ma. It may have been a general change in the global oceanographic system or simply a missing adaptation.

Today its dominance can be attributed to a broad tolerance to a wide range of environmental conditions and to extremely high cyst production rates with dominance of a motile cyst stage (Tangen et al., 1982; Dale, 1992b). It may be possible that the cyst-theca relationship of *T. heimii* changed over the course of time and that, in opposition to today's dominance of the cyst stage, in the past the thecae stage dominated and only very few fossilisable cysts were produced.

<u>*C. levantinum: cool - mesotrophic – eutrophic - DCM*</u>

C. levantinum today is one of the most common calcareous dinoflagellate species in outer neritic and oceanic regions of the south and equatorial Atlantic Ocean (Vink, 2004), as well as of the Mediterranean Sea, with highest percentages in the western part (Meier and Willems, 2003). Due to its adaptation to a wide range of environmental factors (SST, salinity and nutrients), it is considered to be an opportunistic species (Vink, 2004). However, in the Mediterranean Sea it shows a clear eastward trend towards decreasing abundance (Meier and Willems, 2003). It follows the main physical gradients of the surface water, i.e. from west to east increasing temperature and salinity correlated with decreasing nutrient levels.

In the Atlantic Ocean highest abundances of *C. levantinum* are usually found in regions where subsurface waters are relatively cool (13 to 20C°), well mixed and nutrient-enriched, with salinities varying between 34 and 36 psu (Vink, 2004). A positive relationship of *C. levantinum* to cold surface waters of shallow marine to deep shelf environments has been also documented from fossil records of the upper Eocene North Atlantic (Denmark)

(Kohring, 1993a). Here *C. levantinum* significantly dominates (67%) the calcareous dinoflagellate cyst association. A partiality of *C. levantinum* for nutrient rich environments is confirmed by a study from the Holocene eastern Mediterranean Sea, for the time of sapropel S1 formation (Meier et al., 2004). During sapropel S1 formation the increased abundances of *C. levantinum* were linked to a shift of the DCM into the photic zone. A positive correlation of *C. levantinum* with the depth of the DCM layer was also suggested by Vink et al. (2000) and Vink (2004). This depth habitat can also explain its high abundances in the otherwise oligotrophic western Mediterranean Sea today (Meier and Willems, 2003).

In the Pliocene eastern Mediterranean (Bison et al., 2009) it occurred for the first time likely due to nutrient enrichment caused by enhanced continental runoff. Therefore, we here use *C. levantinum* as an indicator for relatively cool waters with elevated nutrient concentrations.

C. stella: warm – mesotrophic - nearshore

So far, *C. stella* cysts were never recorded from modern oceans. However, in a trap sample from the western tropical Atlantic Ocean, Dale (1992) describes some dinoflagellate cysts as *Thoracosphaera albatrosiana* "granular form" sensu Fütterer, 1977, which actually may be attributed to *C. stella* (Bison et al., 2007; 2009). This supports the assumption that this species still exists in modern oceans. The occurrence in the tropical realm of the Atlantic Ocean agrees well with its fossil-inferred preference for warm to temperate mesotrophic and more shelfward environments with normal or even slightly increased salinities (Keupp and Kohring, 1999; Streng et al., 2002, 2004; Bison et al., 2007, 2009).

Fossil *C. stella* cysts until now are only described from low and middle latitudes associated with warm to temperate environments inside the Mediterranean (Keupp and Versteegh, 1989; Kohring, 1993a, b, 1997; Keupp and Kohring, 1999; Bison et al., 2007; 2009; this work) and outside, from the Atlantic (Fütterer, 1977, as Thoracosphaera sp. 2; Hildebrand-Habel and Willems, 2000) and the Indian Ocean realm (Streng et al., 2002, 2004). In the past Indian Ocean, highest concentrations of *C. stella* were found in Middle Miocene sediments. This was followed by a continuous decrease until its absence in the upper Quaternary Indian Ocean sediments (Streng et al., 2002). This progressive decrease is related to climate cooling.

Thus, based on its scarce and questionable recent findings in the tropical western Atlantic Ocean and information gained from fossil occurrences, this species is related to relative warm, mesotrophic waters (Kohring, 1993a, b; Keupp and Kohring, 1993; Streng et al., 2002; Bison et al., 2007, 2009). Due to the lack of information from their living counterparts it is difficult to characterise its environmental preferences more exactly.

P. parva: warm - mesotrophic – nearshore - neritic

P. parva has neither been described from modern oceans nor from the recent Mediterranean Sea. Consequently only little is known about its environmental preferences. The most recent occurrences of *P. parva* are documented from the early Pleistocene southeast Atlantic Ocean where it shows a clear preference for shallow waters (Fütterer, 1984). In the Neogene eastern Mediterranean (Crete) higher numbers of *P. parva* are correlated with increased frequencies of ascidien sclerites (Frydas and Keupp, 1992), which are indicative for neritic environments (Fütterer, 1977). A preference of *P. parva* for nearshore and warm-water environments has also been suggested by Hildebrand-Habel and Willems (2000) and by Hildebrand-Habel and Streng (2003) from studies of the past South Atlantic Ocean. In the upper Miocene and early Pliocene eastern Mediterranean (Bison et al., 2007, 2009) *P. parva* also reflects neritic environments. Here we use it as well as an indicator for nearshore and neritic environments with somewhat warmer surface waters and moderate nutrient levels.

L. urania: halophytic – seasonal - oligotrophic

A restriction of *L. urania* to the highly oligotrophic eastern Mediterranean Sea where it usually dominates the dinoflagellate community is obtained from recent to sub-recent surface sediments by Meier and Willems (2003). Hitherto *L. urania* has never been described from the Atlantic Ocean, neither in the current nor in the past one. Bison et al. (2007, 2009) therefore suggest that it probably survived the MSC in some remnant basins of the Mediterranean itself. So far, fossil representatives of *L. urania* are only described from the Mediterranean (Kohring, 1997; Zonneveld et al., 2001; Meier et al.; 2004; Bison et al., 2007; 2009; this work) and Indo-Pacific realm (Streng et al., 2004).

According to its recent to sub-recent distribution in the Mediterranean Sea *L. urania* seems to be well adapted to a broad temperature range, low nutrient levels and high salinities. Meier and Willems (2003) propose *L. urania* as a halophytic species. Its almost complete restriction to the eastern Mediterranean Sea suggests a narrow tolerance to salinity variation and may reflect a stenohaline character. If this species entered the Mediterranean from the east it should also be present in Neogene sediments from the Indo-Pacific area but not necessarily in those of the Atlantic area. In fact, this is the case (Streng et al., 2004). As this species obviously had its first Pliocene appearance somewhat earlier in the eastern Mediterranean Pissouri Basin (see Bison et al. 2007; 2009) than in the more western part of the Caltanissetta Basin (this study), it may have been reintroduced from the east or survived in marine refuges as aforementioned. Eventually here we use *L. urania* as a representative for highly saline, oligotrophic surface water conditions adapted to high seasonal temperature variations.

Pernambugia tuberosa: oligotrophic - well mixed - offshore

Recent representatives of *P. tuberosa* have only been documented from the Atlantic Ocean so far (Vink, 2004). In the present Mediterranean Sea it is missing. *P. tuberosa* is one of the most common and widespread calcareous dinoflagellate species in the tropical and subtropical Atlantic Ocean today (Vink, 2004). Herein highest abundances were recorded in the well mixed surface layers of relatively deep ocean areas (3000-4000m). Although *P. tuberosa* apparently prefers deep water environments, it has also been recorded, but in minor amounts, in shallower waters up to neritic settings (Vink, 2004).

After Vink (2004), *P. tuberosa* is a species which definitely prefers oligotrophic surface waters, whereas temperature does not significantly affect its distribution. The fact that this species so far was not found in the highly productive recent Arabian Sea (Wendler et al., 2002a) encourages this assumption. Conversely its absence in the highly oligotrophic present Mediterranean Sea seems to be contradictory to this, with the conclusion that environmental features other than eutrophication must be also adverse for *P. tuberosa*.

The present Mediterranean Sea is highly stratified most of the year and only mixed during the winter season. A well mixed surface layer induces turbulence within the upper water column, supporting the transport of nutrients from deeper levels to the photic zone and avoiding the vertical sinking of light limited phytoplankton out of the photic zone. *P. tuberosa* has a relatively compact cyst body, most likely heavier than water. Thus, the mixing may help to keep it in the photic zone. This may also explain *P. tuberosa's* absence in our upper Messinian assemblages, where progressive restriction of the basin was followed by increased stagnation and consequently decreased mixing processes of the surface layer. Other unfavourable factors for *P. tuberosa* in the Mediterranean Sea today may be high salinity and strong temperature seasonality. In the Mediterranean, youngest occurrences of this species were documented from the Late Miocene/Early Pliocene (Kohring, 1993; Bison et al. 2007; 2009; this work). In our study we use *P. tuberosa* as a representative for highly oligotrophic and well mixed conditions of deeper settings. In comparison to *C. albatrosianum* it probably has a higher temperature tolerance but also prefers low seasonal temperature contrast (Vink, 2004). Compared with *L. urania* (~39 psu) it prefers salinities between 36 and 37 psu (Vink, 2004).

Rhabdothorax spp.: Eutrophic – cool – coastal - neritic

The genus *Rhabdothorax* is geographically widespread and known since the Early Cretaceous (Keupp, 1987). *Rhabdothorax* spp. is a cyst forming species complex, characterising a collection of spiny cyst morphotypes which mainly represent species of *Scrippsiella trochoidea*. Some of these morphotypes might also belong to *Scrippsiella regalis*, which is primarily associated with cool open oceanic environments (e.g. Vink et al., 2000; Meier and Willems, 2003; Vink, 2004). *S. trochoidea* on the other hand is associated with shelfward and neritic environments characterised by low and variable temperatures, mixed surface waters and high nutrient concentrations (e.g. Zonneveld et al., 1999; 2000; Janofske, 2000; Wendler et al., 2002a; Vink, 2004). A clear offshore/onshore signal of our *Rhabdothorax* spp. record may therefore be blurred. A broad tolerance to fluctuating environmental parameters, in particular to water turbulence and temperature was derived from its affinity for active upwelling areas (Wendler at al., 2002a). Furthermore a distinct positive relationship between *C. levantinum* and *Rhabdothorax* sp. (*S. trochoidea*) has been reported by Vink (2004).

In the modern Mediterranean Sea species of *Rhabdothorax* spp. have been found in surface sediments of shallower and more coastal regions, with highest occurrences in the Gulf of Lion (Meier and Willems, 2003) and in the Gulf of Naples (Montresor et al., 1998) because both settings are strongly influenced by eutrophication through river runoff. To conclude, high frequencies of this species we use as an indicator for relative cool, nearshore and neritic environments with higher nutrient levels.

Environmental evolution

Based on the distribution pattern of the calcareous dinoflagellate cyst assemblages investigated in this work, we defined environmental changes of the Caltanissetta Basin in seven more or less distinct phases. For the pre-evaporitic upper Tortonian/Messinian we distinguished three major environmental changes, leading from relatively stable, predominantly oligotrophic and warm waters to very unstable water conditions with fluctuating temperature and nutrient levels towards the MSC. For the early Pliocene sequence, directly following the MSC, our data reflect four stages of environmental change towards a more modern Mediterranean system, similar but not yet identical with the present one. In the following sections, the calcareous dinoflagellate cyst distribution and their environmental relationships will be discussed.

Tortonian/Messinian succession

Interval 1: oligotrophic – instable – warm to cool

During the upper Tortonian, the almost exclusive presence of oligotrophic species indicates strongly nutrient depleted surface waters. This correlates with the interpretation of the eastern Mediterranean basin during this time by Kouwenhoven et al. (2006), who proposed subtropical oligotrophic surface water conditions. Within our study, dominance shifts of the main species *C. albatrosianum*, *P. tuberosa* and *L. urania* in addition reflect fluctuations in temperature, water turbulence and salinity of the continuously extreme oligotrophic surface waters. A first expression of a notable environmental change is indicated by the significant drop in total cyst concentration at the top of the Tortonian (at ~7.51 Ma) already about 1.5 million years before the onset of the MSC (5.96 Ma). We interpret the associated drop of *C. albatrosianum* as a short lasting cold-water impact. This early expression of Mediterranean Basin restriction correlates with partial uplift of the Rif Corridor (Krijgsman et al., 1999b) and the Melilla Basin (Van Assen et al., 2006). The resulting restriction of the Mediterranean/Atlantic water exchange may have caused siphoning of cold Atlantic waters into the Mediterranean basin (Van Assen et al., 2006).

Interval 2: stable - warm - oligotrophic

During the subsequent lower Messinian the persisting dominance of *C. albatrosianum* indicates relatively stable warm and oligotrophic surface waters up to about 7.17 Ma when a sudden drop of *C. albatrosianum* and a concurrent increase of more neritic and nutrient preferring species (*P. parva*, *Rhabdothorax* spp., *L. granifera*), suggests a further expression of early basin restriction long before the real MSC began. At this point, we suggest a shallowing of the Caltanissetta Basin, associated with a slight increase in the nutrient levels and somewhat lower temperatures. Additionally, the low percentage of the halophytic species *L. urania* reflects lower salinities, which may have resulted from increased continental runoff probably associated with increased transport of neritic waters into the Caltanissetta Basin.

Interval 3: instable – warm to cool - mesotrophic to oligotrophic

After the predominantly warm and oligotrophic upper Tortonian/lower Messinian, the further reduction in abundance of *C. albatrosianum* at ~6.8 Ma indicates the next environmental shift; a significant drop in sea surface temperature. From this moment, up to the onset of the evaporite formation, the cumulative decrease in total calcareous dinoflagellate cyst numbers in combination with highly fluctuating relative and absolute abundances indicate a period with extremely unstable surface water conditions. This includes frequent and high amplitude shifts in the surface water, particularly in temperature and trophic state. Conversely, this highly variable environment obviously facilitated the co-existence of different species and thus maintains a high diversity.

Species adapted to higher nutrient level and lower temperatures, such as *L. granifera* and *T. heimii*, intermittently notably increased suggesting an even more restricted environment. This interpretation is in agreement with the gradual separation of the Mediterranean basin from the Atlantic Ocean, caused by intensified tectonic activity in the Gibraltar area during this period (e.g. Ryan, 1973; Blanc, 2000; Krijgsman et al., 1999b; Duggen et al, 2003; Krijgsman and Garces, 2004; Meijer and Krijgsman, 2005; Van Assen et al., 2006). As a consequence, progressive instability and restriction of the Mediterranean basin led to increased inhospitable environmental conditions towards the upper Messinian (e.g. Krijgsman et al., 1999b; Kouwenhoven et al., 2006; Van Assen et al. 2006). However, our data do not support an outstanding shallowing of the Caltanissetta Basin during this stage. Species indicative for shallower waters (e.g. *P. parva*) almost completely disappeared during this upper Messinian stage. A solution to this paradox may be found in the dinoflagellate life cycle of neritic species as they rest on the sea floor. Inhospitable bottom waters, probably due to early brine formation or due to stagnant conditions (see Kouwenhoven et al., 2006) may have interrupted this life cycle, resulting in the disappearance of neritic taxa. Simultaneously, the open oceanic species *T. heimii* and *C. albatrosianum* were greatly increased at times, suggesting periods of amplified Atlantic water inflow, marked by different temperature and nutrient regimes. The highest extent of this environmental progress towards increasingly unfavourable conditions in the Caltanissetta Basin was reached during the uppermost Messinian, at about 5.97 Ma. Here, samples barren of calcareous dinoflagellate cysts indicate their disappearance, shortly before the onset of the MSC.

Early Pliocene succession

With the beginning of the Pliocene epoch (5.33 Ma) almost all species present before the major restriction of the Mediterranean basin reappeared immediately or after a short delay. Additionally, we observed the first occurrence of *C. levantinum*, a proposed Atlantic newcomer (Bison et al., 2009). However, the reappearance of most taxa did not lead to a reestablishment of the Miocene assemblages. The dinoflagellate cyst community significantly changed. *C. albatrosianum*, the dominant species of the Tortonian/Messinian period, is replaced by *L. granifera*. This turnover in the dinoflagellate cyst community probably marks a major change in the hydrological and climatic conditions of the Mediterranean Basin. The predominantly warm and oligotrophic environment of the upper Tortonian/lower Messinian changed into a somewhat cooler and/or more seasonal, but also nutrient enhanced phase, culminating with an outstanding eutrophication signal (*L. granifera* peak) just above the Miocene-Pliocene boundary. Our early Pliocene sequence actually represents a four stage transitional phase towards a more modern Mediterranean system, similar but not yet identical with the present one.

Interval 4 (5.33 Ma): eutrophic - cool

Our Pliocene sequence starts with a prominent abundance peak of *L. granifera*, indicating high nutrient concentrations in the upper photic zone, probably associated with lower salinities. This further is confirmed by the absence of the halophytic species *L. urania*. Minor abundance peaks of *C. levantinum* and *Rhabdothorax* spp., both indicative for enhanced nutrient levels, emphasise this early Pliocene eutrophication event. The warm water species *C. albatrosianum* is almost missing, which indicates cooler surface waters. The high nutrient content in the surface waters we relate to a strong fluvial phase coinciding with enhanced continental runoff. This phase obviously already started during the late Messinian "Lago Mare" stage and continued up into the Early Pliocene epoch (Griffin, 2002). We furthermore suggest that continuing transgression of denser Atlantic water into the earliest Pliocene Mediterranean basin, previously filled with the less dense water of the Lago Mare period, might have induced an estuarine upwelling system, which in addition periodically delivered nutrient rich waters to the photic zone (see Bison et al., 2009).

A similar environmental situation, also with an abundance peak of *L. granifera*, just above the Miocene-Pliocene boundary, is described in the first study of this project from the eastern Mediterranean Pissouri Basin (Bison et al., 2007; 2009). Here, the Early Pliocene is interpreted as a period of increased freshwater runoff, associated with local upwelling processes, the latter caused by a temporal reversal of the circulation pattern towards an estuarine system. As a consequence, eutrophication of the photic zone finally led to increased primary production in the calcareous dinoflagellate cyst community, mainly of *L. granifera* (see Bison et al., 2007; 2009). The same scenario is proposed for our Sicilian succession.

Interval 5: mesotrophic - warm to moderate

After the initial nutrient-rich phase this interval starts with an abrupt drop in total abundance of *L. granifera*, indicating a considerable decrease in the nutrient concentrations in the surface water. At the same time, increased numbers of *C. albatrosianum* and *C. stella* reveal a rise in water temperature with the highest values in the upper part of the interval. Concurrently, several late Miocene species re-appeared suggesting less extreme surface water conditions. We think that a drastic change in the hydrological system is responsible for this shift, which is from an initially estuarine to increasingly anti-estuarine circulation, most likely in combination with a change to a somewhat drier and warmer climate period. As a consequence, a pronounced loss of nutrients introduced into the surface waters through both, upwelling and riverine runoff, probably took place. The presence of the halophytic species *L. urania* indicates slightly increased salinities, likely caused by somewhat higher evaporation rates. Overall, we propose an environment marked by mesotrophic surface waters with low to moderate temperatures and normal to slightly increased salinities.

Interval 6: mesotrophic to eutrophic – moderate to cool

Here, a notable rise in *L. granifera* abundances indicates a second eutrophication event in the surface waters with highest nutrient levels at the beginning of this phase. A general decrease of *C. albatrosianum* attenuated by repeated abundance shifts reveals lower and varying water temperatures. We attribute this situation to a change toward a more humid and somewhat cooler climate phase, again coinciding with increased continental runoff. In addition a slight increase in the percentage of the deep dwelling species *T. heimii* suggests the establishment of a more stable water column pointing to a relaxation of the hydrological situation in this area. Conversely, a general increase of *P. parva* with a distinct abundance peak in the middle part of this interval indicates at least a temporay shallowing of the basin, which may be linked to local uplift processes. This is supported by a drop of *T. heimii*, contemporaneous with the abundance peak of *P. parva*. In summation, we suggest environmental conditions characterised by alternating sea surface temperatures, salinity and water depth and relatively high nutrient concentration.

Interval 7: warm to moderate – oligotrophic to mesotrophic

A further significant change in the Pliocene dinoflagellate cyst assemblage terminates our sequence. The hitherto dominating eutrophic species *L. granifera* is replaced by *C. albatrosianum*. The main cause of this turnover is, we suggest, a drastic change in the nutrient budget of the surface water linked to a drier and warmer climate period. As a consequence, reduced continental runoff most likely led to a nutrient loss in the surface waters. This finally may have led to a modification of the circulation pattern towards an anti-estuarine system, which caused an additional nutrient loss in the surface waters through downwelling processes. Warming and relaxation of the surface water due to the failure of upwelled cold waters were likely a further consequence of such a reversal. Interestingly *C. levantinum*, the Atlantic newcomer, slightly increased whereas *T. heimii* strongly decreased again. The decrease of *T. heimii* is linked to the suggested strong nutrient loss most likely also affecting the deeper layer. The simultaneous increase of *C. levantinum* we explain by a stronger Atlantic influence and a better adaptation of this opportunistic species to lower nutrient levels.

Comparison of our final Pliocene and present Mediterranean dinoflagellate cyst community

Regardless of these progressive changes, our final Pliocene dinoflagellate cyst association still differs from that of the present Mediterranean Sea (see Bison et al., 2007) suggesting substantial differences in the hydrological and/or climatic conditions. Temperatures are most likely still warmer and less seasonal than today (dominance of *C. albatrosianum*) and salinities probably did not reach its present high level yet (low percentages of *L. urania*). Nutrient concentrations in the surface water are comparable to those of nowadays (similar *L. granifera* concentrations).

In the present Mediterranean Sea *C. albatrosianum* is very scarce and does not reach more than 5% (Meier and Willems, 2003). Here, the most adverse environmental factor for *C. albatrosianum* seems to be the strong seasonal contrast in sea surface temperature. The low abundance of *L. granifera* today (> 1%) and in our uppermost Pliocene assemblage (9%) is related to low nutrient availability in the surface water due to seasonal downwelling processes (anti-estuarine circulation). Deviations in the two systems can be explained by differences in climatic and hydrological conditions. Today, as in the past, the Mediterranean Sea is and was influenced by different climate zones.

The climate of the Early Pliocene is thought to have been warm and dry in the south-western Mediterranean region and warm and humid with higher annual precipitation rates along the Adriatic coast (Fauquette et al., 1998; Kovar-Eder et al., 2006). Rather cool temperatures and increased precipitation was documented for the end of the Messinian stage in the Adriatic Sea (Fauquette et al., 2006). This humid phase possibly extended into the earliest Pliocene and thus supports our findings of somewhat cooler and humid conditions just at the beginning of the Pliocene epoch. Even though local climatic conditions affected the

Mediterranean basins, a number of changes in the dinoflagellate cyst association are instead caused by hydrological variations (e.g. circulation pattern, warm/cold currents) than by climate. Tectonic activity in the Gibraltar region led to the restriction of the Atlantic-Mediterranean water exchange and finally to the MSC. Additionally, local tectonic activity affected the hydrological situation of the different basins. Today the Mediterranean Sea is divided into two main basins, western and eastern, which are separated by a physical barrier in the Strait of Sicily. This separation goes along with a west-east gradient in the main physical properties (salinity increase, temperature increase, nutrient decrease).

Accordingly, the dinoflagellate association of in the present Mediterranean Sea can be differentiated into a western and eastern one. Both basins have in common that *T. heimii* clearly dominates the dinoflagellate community with 88% in the eastern and 99% in the western basin (Meier and Willems, 2003). The lower percentages in the eastern basin are explained by higher absolute abundances of the other cysts and not by a decrease in productivity of this species (Meier and Willems, 2003). Apart from this, the main species of the western basin is the opportunistic species *C. levantinum*, followed by the open ocean species *Scripsiella regalis* (Meier and Willems, 2003). In the eastern basin the halophytic species *L. urania* clearly dominates followed by *C. elongatum* (Meier and Willems, 2003).

Our current and previous (Bison et al., 2007, 2009) results show that this east-west separation and establishment of the modern Mediterranean calcareous dinoflagellate association must have taken place after the earliest Pliocene stage. In our Sicily record the aforementioned most abundant Mediterranean species quantitatively only play a subordinated role and *C. elongatum* is even missing. This species may have evolutionary separated from *C. levantinum*, as both show broad similarities in outer shape and in ultrastructure (Meier et al., 2002). Interestingly, this species was documented in the eastern Mediterranean Pissouri Basin during this time as an Atlantic newcomer (see Bison et al., 2009).

T. heimii, the presently dominating calcareous dinoflagellate species, is rather rare in our Sicily record. The same holds for the eastern Mediterranean basin during that time (Bison et al. 2007; 2009). It becomes more frequent and a constant component in both basins with the onset of the Pliocene. Actually, in our Sicily record, the first expressions of its later dominance became evident with its distinct abundance peaks already during the upper Messinian. These peaks are most likely associated with temporalily enhanced inflow of Atlantic waters, facilitating the establishment of a DCM layer. As *T. heimii* today prefers relatively cold water levels and possibly did so during the Miocene/Pliocene, similar to *C. levantinum*, it may have been simply too warm for these species. *T. heimii*, in addition, may have been hampered by insufficient stability of the upper water column due to ongoing reorganisation processes of the freshly refilled Mediterranean basin. For *L. urania*, salinity may have been too low, probably due to stronger continental dilution of the surface water through enhanced continental runoff. Anyway, during that time the environmental conditions must have been still unfavourable for the present Mediterranean species. It was probably a combination of various environmental and evolutionary circumstances limiting their distribution.

Marl/sapropel related species distribution pattern

Marl/sapropel layers in the Mediterranean realm are generally related to precession controlled climatic and oceanographic changes, driven by variations in solar radiation (e.g. Rohling and Hilgen, 1991; De Lange et al., 2008). Sapropels, enriched in organic carbon, are closely linked to humid climates during precession minima and insolation maxima (Rohling and Hilgen, 1991). At these times, increased freshwater runoff, associated with amplified nutrient input into the Mediterranean basin occurred due to enhanced summer precipitation (e.g. Rossignol-Strick et al., 1983; Rossignol-Strick, 1985; Hilgen, 1987; 1991; Rohling and Hilgen, 1991; De Lange et al., 2008). As a consequence decreased surface water salinity usually lead to enhanced stability of the water column and the establishment of an estuarine circulation (e.g. Rossignol-Strick et al., 1982; Rohling and Hilgen, 1991; Fontugne and Calvert, 1992) which stimulates local upwelling processes (Passier et al., 1999). Although most sapropels formed during humid climate phases (e.g. Rossignol-Strick et al., 1982; Rossignol-Strick, 1985; Cramp et al., 1988; Arnaboldy and Meyers, 2006) a simultaneous climate warming is not obligatory, as sapropels have developed during glacials as well as during interglacials (Rohling and Hilgen, 1991). The same seems to be valid for the sea surface temperature during sapropel times. This at least is revealed by our dinoflagellate cyst record, suggesting that sapropels were formed under different environmental conditions at certain times. Although the overall mechanisms of the sapropel forming process are widely well-known, there are still discrepancies related to the manifold environmental and hydrological conditions during times of sapropel formation.

Within our record from Sicily no clear and continuous relationship between marl/sapropel layers and individual dinoflagellate cyst species could be distinguished. However, correlations between increased cyst concentrations and sapropel layers were observed in the upper Messinian. In the lower Messinian, this correlation could not be observed it even seems to have been reversed. The same holds for the upper Tortonian and

Pliocene record. In the upper Messinian different species, such as *C. albatrosianum*, *T. heimii* and to a minor part *L. granifera*, peaked during times of sapropel formation, reflecting different environmental conditions at such times. For the times of sapropel formation we differentiated at least three different environmental conditions in the surface waters.

1. *C. albatrosianum* dominated sapropels: Warm and oligotrophic surface water conditions prevailed with a well stratified water column.
2. *T. heimii* dominated sapropels: Cool and stratified waters prevailed, with a nutrient depleted upper photic zone and high nutrient levels at depth, establishing the formation of a DCM layer.
3. *L. granifera* dominated sapropels: High nutrient concentrations, low temperatures and salinities prevailed in the upper surface water.

Sapropel formation is usually associated with warm and humid climates in the Northern Hemisphere – generally accompanied by enhanced precipitation and runoff and thus eutrophication of the surface waters – but our results show that this association is simplified and not always valid for the conditions in the surface waters. Our results show that greater productivity does not always result in sapropel formation and for the formation of sapropels a considerable increase in productivity is not required if water circulation is restricted. On the other hand, we have shown that sapropels form as well under oligotrophic and stratified surface water conditions.

Comparison Caltanissetta Basin/Pissouri Basin

The dinoflagellate cyst record from the Caltanissetta Basin has been compared with that of the Pissouri Basin (Bison et al., 2007, 2009). Overall, the calcareous dinoflagellate cyst distribution pattern of the Caltanissetta Basin resembles that of the Pissouri Basin during Late Miocene/Early Pliocene time. Broadly similar major trends were detected with respect to the species composition and distribution preceding and following the MSC. Hence, our results from the Caltanissetta Basin confirm those obtained from the Pissouri Basin (Bison et al., 2007; 2009). Our data confirm that the environmental changes caused by the MSC and the establishment of normal marine conditions during the earliest Pliocene occurred more or less simultaneously within the central and eastern Mediterranean basin.

Overall, both settings experienced a roughly similar environmental evolution. It started with a relatively stable warm and highly oligotrophic upper Tortonian/lower Messinian time (*C. albatrosianum* dominance) towards a very instable and increasingly restricted environment during the upper Messinian (dominance shifts of *C. albatrosianum*, *T. heimii*, *L.*

granifera), through to a nutrient-enriched but somewhat cooler Pliocene period (*L. granifera* dominance), which finally ended with the establishment of again somewhat warmer and nutrient depleted surface water conditions (*C. albatrosianum* dominance).

Nevertheless, although the overall environmental evolution of the two basins appears to be comparable across the studied stratigraphic interval, some minor differences in both, species distribution and species composition exist. The Pliocene Atlantic newcomer *C. elongatum* and some rare species found in the Pissouri Basin (*B. tricarinelloides* and *P. loeblichii*) could not be detected in the Caltanissetta Basin. On the other hand some rare species, present in the Caltanissetta Basin (e.g. *P. sicelis*, *P. rhombica* and *P. polymorphum*) are missing in the Pissouri Basin. Cyst concentrations are usually significantly higher in the Pissouri Basin record, especially with respect to the main species *C. albatrosianum*, *L. granifera* and *C. stella*. The exceptional abundance peaks of *T. heimii* and *C. albatrosianum* during the upper Messinian are missing in the Pissouri Basin, or could not be detected because the samples are not really time equivalent. The first notably increased accounts of *T. heimii* in the Pissouri Basin were recorded at the end of our Pliocene.

The termination of the Miocene calcareous dinoflagellate cyst association shortly before the onset of the MSC occurred later in the Pissouri Basin than in the Caltanissetta Basin, at least later than 5.93 Ma (see Bison et al., 2007). Species such as *C. albatrosianum*, *C. stella* and *L. granifera* - although highly diminished - are still present in the Pissouri Basin during this late Messinian time. The co-occurrence of these different species and their strong reduction though, suggest already strongly instable and restricted environmental conditions. In addition, relatively high percentages of undefined species point to some modification of the cyst shape most likely due to diagenetic overprint and reworking.

The final drop of *L granifera* during the Pliocene in the Caltanissetta Basin is associated with a distinct drop of *T. heimii*. Concurrently *C. albatrosianum* increased (both, in absolute and relative abundances) together with some rare species of neritic origin. Here, the essential decrease in the trophic level (*L. granifera* drop) was obviously associated with a warming of the surface water and may be accompanied by a stronger continental influence or shallowing, resulting in an increase of neritic species. In contrast to this, the final drop of *L. granifera* in the Pissouri Basin is concurrent with a notable increase of *T. heimii* and a decrease of *C. albatrosianum* in relative abundance. It is accompanied by a minor increase of *C. levantinum* and *C. elongatum*, indicating a further step towards more modern Mediterranean conditions. The relative decrease of *C. albatrosianum* at this point is actually related to an absolute increase of *T. heimii* and consequently does not reflect a significant decrease of the surface water temperature. Nevertheless, in the eastern Mediterranean basin increased *T. heimii* numbers may indicate the establishment of a DCM layer, which - at least at the deeper photic zone - is associated with higher nutrient concentrations and lower temperatures. These differences in the calcareous dinoflagellate cyst records of the two basins may be caused by several factors such as oceanography (circulation), regional tectonics (uplift), influences of the surrounding borderlands (continental runoff) and different climatic regimes (Asian Monsoon/North Atlantic/Northern Borderlands). Nevertheless, the general trends in the environmental evolution, such as the transition from normal marine to restricted conditions and the re-organisation of the marine settings, must have taken place more or less simultaneously in the two basins, as reflected in the dinoflagellate cyst records.

Summary

Changes in diversity, species richness, absolute and relative abundance have been analysed and interpreted in terms of the palaeoenvironment. Calcareous dinoflagellate cyst distribution pattern were used to infer changes in the environmental evolution of the central Mediterranean Caltanissetta Basin, prior to and immediately after the MSC. The environmental changes are nicely reflected in the calcareous dinoflagellate cyst distribution pattern. We distinguished seven major environmental phases based on variations in the calcareous dinoflagellate cyst associations. Three of them we assigned to the pre-evaporitic upper Tortonian/Messinian time and four to the early Pliocene sequence.

- Interval 1: Highly oligotrophic conditions with fluctuations in SST and salinity. The representative main species are *C. albatrosianum*, *P. tuberosa* and *L. urania*.
- Interval 2: Stable warm and oligotrophic conditions, represented by the continuous dominance of *C. albatrosianum*.
- Interval 3: Unstable conditions with varying SST and nutrient levels, indicated by distinct shifts in species dominance; a general decrease in the cyst concentrations reflect progressive restriction.
- Interval 4: Cool eutrophic conditions, indicated by a prominent abundance peak of *L. granifera,* which distinctly dominates.
- Interval 5: Mesotrophic conditions with moderate SST. The main species is *L. granifera* followed by *C. albatrosianum* and *C. stella*.

- Interval 6: Mesotrophic to eutrophic conditions with moderate to cool SST. *L. granifera* dominates followed by *P. parva*, *C. albatrosianum* and *T. heimii*.
- Interval 7: Oligotrophic to mesotrophic conditions with moderate to warm SST. *C. albatrosianum* dominates, followed by *P. parva*, *Rhabdothorax* spp. and *L. granifera*, the latter one significantly decreased.

The results of our study show that the alteration in the nutrient level and in SST are the main physical factors affecting the calcareous dinoflagellate cyst species distribution during the studied interval. Most of the Mediterranean species probably re-entered the Mediterranean from the Atlantic during the early Pliocene. In contrast *L. urania*, which today dominates the eastern Mediterranean assemblages, probably re-entered the Mediterranean from the east, or survived the MSC in some marine refuges of the Mediterranean itself. Various rare neritic species occurred for the first time within the upper Pliocene samples and may indicate a shallowing shortly after its initial refilling or increased influence of neritic waters. *C. levantinum*, a common representative of the current western Mediterranean Sea and the Atlantic Ocean, occurred for the first time within the early Pliocene record. This species can be seen as an Atlantic newcomer and it indicates somewhat cooler temperatures or a stronger seasonal contrast in sea surface temperature.

A general turnover in the calcareous dinoflagellate cyst community took place with the beginning of the Pliocene epoch. Here, *C. albatrosianum*, the dominant species of the Tortonian/Messinian assemblages, has been replaced by *L. granifera*, the prevailing species of the early Pliocene. This major turnover reflects a change in the surface water of the Caltanissetta Basin from primarily warm and oligotrophic to somehow cooler, eutrophic to mesotrophic conditions. At the top of our Pliocene succession a period of nutrient depleted and warmer temperatures again becomes apparent (recovery of *C. albatrosianum*).

Overall our dinoflagellate cyst record shows a notable decline in the mean surface water temperature with the beginning of the Pliocene as indicated by the significant decrease of *C. albatrosianum*. Simultaneously a general increase in the nutrient level took place, culminating just above the Miocene-Pliocene boundary as indicated by the prominent peak of *L. granifera*. The decline of the mean sea surface temperature may be seen as a result of a stronger seasonality in the Mediterranean basin. Increased nutrient availability implies a stronger continental influence with enhanced river runoff, probably due to a more humid climate. For the earliest Pliocene phase, we additionally suggest a temporal reversal of the thermohaline circulation system to an estuarine one. This probably led to local upwelling processes and thus enhanced nutrient availability in the surface waters.

Our results confirm those from the eastern Mediterranean Pissouri Basin on Cyprus, suggesting a Mediterranean-wide more or less similar environmental evolution apart from some local differences related to the MSC. When and how the modern Mediterranean calcareous dinoflagellate association were exactly established remains unclear but may be the subject of future studies. Our findings in the cyst distribution pattern confirm the usability of calcareous dinoflagellates for palaeoenvironmental reconstruction.

Acknowledgements

We are grateful to Frits Hilgen and Wout Krijgsman for providing the sample material. Gerard Versteegh is thanked for critical reading and helpful comments and suggestions for a preliminary version of this manuscript. We thank Kara Bogus for final English corrections and suggestions. All members of the working group of Historical Geology and Palaeontology of the University of Bremen are thanked for their general support and openness to discussion. We gratefully acknowledge financial support through the German Research Foundation (DFG) (ProjectWI-725/19-1/2).

Plates

Two plates, composed of 44 scanning electron microscope (SEM) photographs, representatively pictured for the 19 species recorded within this study. For some species morphological variations are shown. The text accompanying the photographs includes the species' name, sample number, number of the SEM-stub, the stratigraphic position and a short description of the specimen illustrated in the figures.

Plate 1 (figs 1 - 24)

Figs 1-5: *Calciodinellum albatrosianum.* **1** (sample 59; SEM-stub It 12/3, lower Pliocene): apical view; closed cyst with polygonal delineated operculum; large pores mark the cyst surface composed of rosette-like arranged wall crystals. **2** (sample 21; SEM-stub FT 20/2, lower Messinian): oblique apical view; open cyst with polygonal archaeopyle. **3** (sample 24; SEM-stub Ft 32/1; upper Messinian): cross section of a broken cyst; granular wall crystals showing a distinct rosette-like arrangement. **4** (sample 6; SEM-stub Gib 57/4; lower Messinian): oblique apical view of an open cyst; pores are reduced due to secondary distal growth of the surface wall crystals. **5** (sample 60; SEM-stub It 14/1; lower Pliocene): lateral apical view; closed cyst with reduced pores due to secondary crystal growth; incomplete suture of the operculum involving only three plates.

Fig 6: *Calciodinellum limbatum.* **6** (sample 5; SEM-stub Gib 59/1; lower Messinian): oblique apical view; open cyst with a polygonal archaeopyle; the single layered cyst wall is formed by radially arranged crystals; small pores are evenly distributed on the cyst surface showing a distinct paratabulation pattern.

Figs 7-9: *Calciodinellum operosum.* **7** (sample 58; SEM-stub It 10/1; lower Pliocene): antapical view; paratabulated cyst with well developed crystal ridges and large pores on the wall surface. **8** (sample 53; SEM-stub It 2/1; lower Pliocene): lateral view; closed cyst with well developed crystallite ridges and paratabulation; pores are reduced. **9** (sample 15; SEM-stub Gi34/2; lower Messinian): lateral apical view; closed cyst with marked suture of the polygonal operculum.

Figs 10-11: *Calcigonellum infula.* **10** (sample 55; SEM-stub It 5/2; lower Pliocene): lateral antapical view, elongated cyst with reduced paratabulation. **11** (the same cyst): zoomed view.

Fig 12-13: *Calciodinellum levantinum.* **12** (sample 54; SEM-stub It 4/3; lower Pliocene): lateral view; open cyst with closely arranged pyramid-like surface crystals. **13** (sample 54; SEM-stub It 4/1; lower Pliocene): closed cyst with relative smooth surface.

Figs 14-16: *Caracomia stella.* **14** (sample 21; SEM-stub FT 20/2; lower Messinian): oblique apical view; spherical cyst with caved polygonal operculum. **15** (sample 55; SEM-stub It 5/2; lower Pliocene): oblique apical view; open cyst with a large polygonal archaeopyle; the single layered cyst wall is formed of radially arranged epitaxially grown crystals. **16** (sample 24; SEM-stub FT 32/4; upper Messinian): cross-section; dehisced cyst showing a highly porous inner cyst surface.

Figs 17-20: *Lebessphaera urania.* **17** (sample 24; SEM-stub FT 32; upper Messinian): top view; closed cyst with relatively large and loosely arranged blocky crystals. **18** (sample 62; SEM-stub It 16/1; lower Pliocene): oblique apical view; cyst with partly caved epitractal operculum. **19** (sample 21; SEM-stub FT 20/2; lower Messinian): oblique apical view; closed cyst with marked suture of the operculum; the cyst surface is composed of very coarse blocky and slightly elongated wall crystals. **20** (sample 21; SEM-stub FT 20/2; lower Messinian): apical view; open cyst with large epitractal archaeopyle.

Figs 21-24: *Leonella granifera.* **21** (sample 40; SEM-stub It 54/2; upper Messinian): oblique apical view; closed spherical cyst with a distinctly marked round operculum; the cyst surface shows a microgranular shape interspersed with numerous small pores. **22** (sample 53; SEM-stub It 2/2; lower Pliocene): oblique apical view; spherical cyst with smooth cyst surface and round archaeopyle showing the single layered cyst wall. **23** (sample 20; SEM-stub FT 11/1; lower Messinian): closed cyst; modified cyst surface due to secondary crystal growth. **24** (sample 9; SEM-stub Gib 50/4; lower Messinian): lateral apical view; open cyst with small round archaeopyle; pores are missing due to secondary growth of the cyst surface wall crystals.

Plate 1

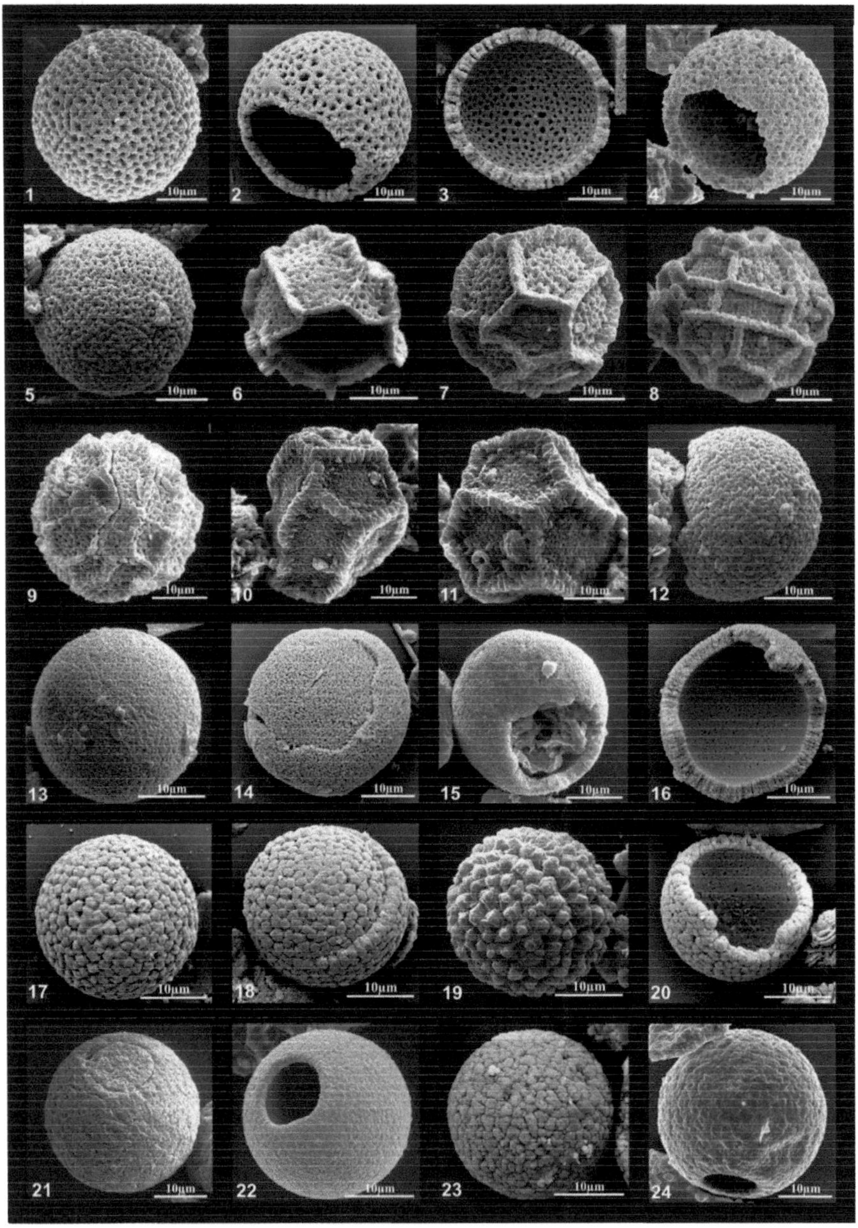
1
10µm
2
10µm
3
10µm
4
10µm
5
10µm
6
10µm
7
10µm
8
10µm
9
10µm
10
10µm
11
10µm
12
10µm
13
10µm
14
10µm
15
10µm
16
10µm
17
10µm
18
10µm
19
10µm
20
10µm
21
10µm
22
10µm
23
10µm
24
10µm

Plate 2 (figs 25-48)

Fig. 25: *Leonella granifera.* **25** (sample 56; SEM-stub It 6/3; lower Pliocene): lateral apical view; closed cyst with irregularly formed pyramid-like surface wall crystals; the suture of the round operculum is visible at the left side.

Figs 26: *Melodomuncula berlinensis.* **26** (sample 62; SEM-stub It 16/1; lower Pliocene): lateral apical view; cylindrical shaped cyst with hexagonal outline, a round archaeopyle is centered on the apical site.

Fig 27: *Pernambugia tuberosa.* **27** (sample 2; SEM-stub Gib 62/1; upper Tortonian): closed spherical cyst; the cyst surface is formed of relatively coarse and closely arranged pyramid-like wall crystals.

Figs 28-29: *Praecalcigonellum polymorphum.* **28** (sample 63; SEM-stub 17/3; lower Pliocene): oblique apical view; small paratabulated cyst with a hexagonal apical outline and a centered round archaeopyle. **29** (the same cyst): lateral antapical perspective.

Figs 30-33: *Pirumella parva.* **30** (sample 60; SEM-stub It 14/3; lower Pliocene): lateral apical view; slightly elongated cyst with a round archaeopyle. **31** (the same cyst): zoomed view; detail of the cyst surface with obliquely arranged rode-like crystals. **32** (sample 60; SEM-stub It 12/1; lower Pliocene): cross section of the double layered cyst wall. **33** (sample 60; SEM-stub It 14/2; lower Pliocene): lateral apical view; spherical cyst with small round archaeopyle.

Fig 34: *Pirumella rhombica.* **34** (sample 58; It 10/3; lower Pliocene): slightly ovoid cyst with flat small disk-like crystals.

Fig 35-37: *Praecalcigonellum schizosaeptum.* **35** (sample 63; It 17/1; lower Pliocene): oblique apical view. **36** (the same cyst): apical view; a rounded vaguely pentagonal outline with a circular operculum in the center. **37** (sample 63; It 17/3; lower Pliocene): apical view; broken cyst displaying a one layered cyst wall with radially arranged crystals.

Fig 38-41: *Rhabdothorax* spp. **38** (sample 60; SEM-stub It 14/1; lower Pliocene): lateral view; spherical closed cyst with regularly and loosely arranged short blocky spines, an epitractal suture of the operculum is visible on the left side. **39** (sample 6; SEM-stub Gib 57/3; lower Messinian): closed cyst with longer spines; the wall crystals are less regular and relative loosely arranged. **40** (sample 21; SEM-stub Ft 20/2; lower Messinian): spherical cyst with relatively long spines arising from the center of loosely arranged irregularly formed basal crystal plates. **41** (sample 2; SEM-stub Gib 62/2; upper Tortonian): lateral view; elongated cyst with more robust spines.

Figs 42-43: *Thoracosphaera heimii.* **42** (sample 24; FT32/1; upper Messinian): oblique apical view; spherical cyst with a small round archaeopyle and irregularly formed surface crystals. **43** (sample; SEM-stub It 10/3; lower Pliocene): lateral view; spherical cyst with a porous cyst surface.

Fig 44: *Ruegenia oranensis.* **44** (sample 2; SEM-stub Gib 62; upper Tortonian): lateral view; spherical cyst with a smooth surface; a suture, marking the outline of the operculum, is weekly visible at the left side.

Fig 45-46: *Pirumella* sp. **45** (sample 59; SEM-stub It 12/2; lower Pliocene): oblique lateral view; paratabulated cyst; the ridges are formed of relative coarse crystals. **46** (the same cyst): detail of the cyst surface.

Fig 47-48: *Pirumella sicelis*. **47** (sample 59; SEM-stub It 12/1; lower Pliocene): spherical cyst with irregular arranged angular surface crystals. **48** (sample 57; SEM-stub It 8/1; lower Pliocene): slightly elongated cyst with highly irregular and loosely arranged surface crystals.

Plate 2

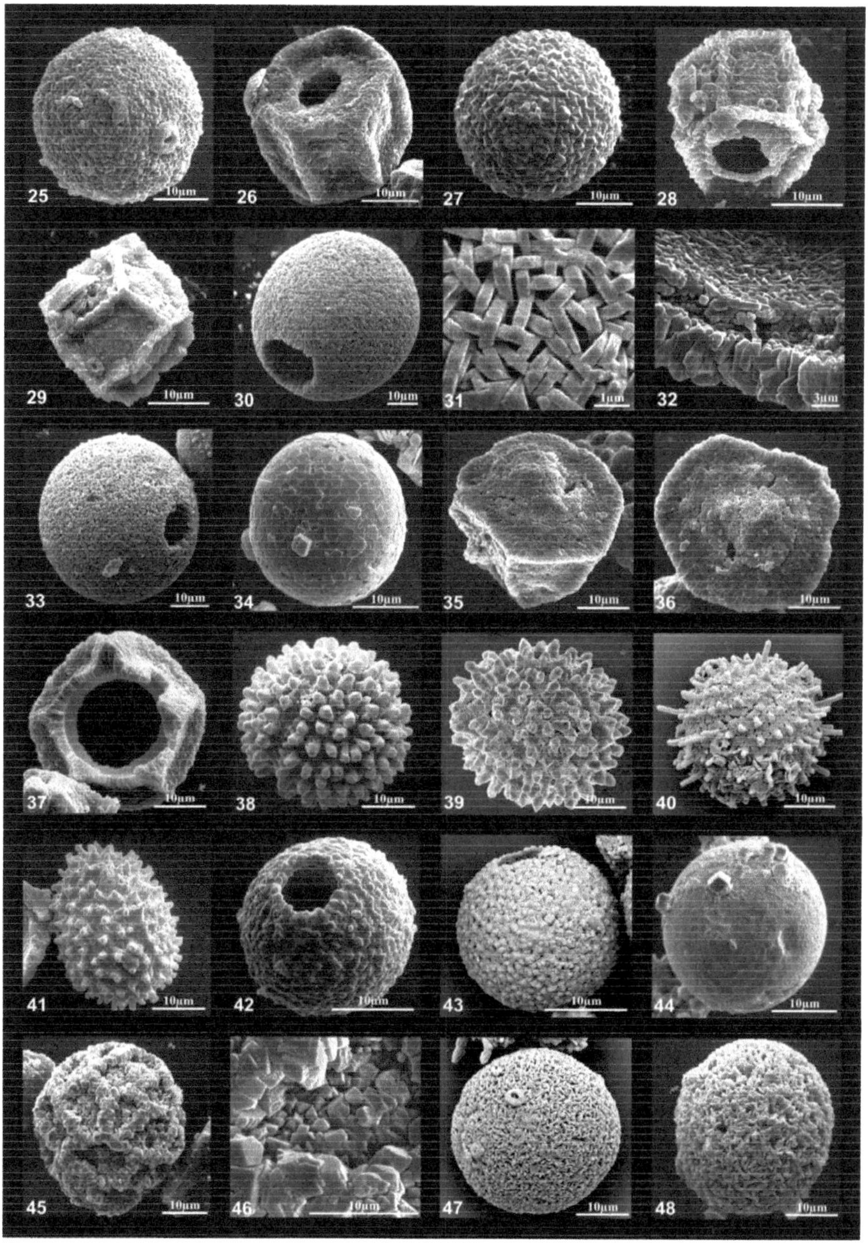
25
10μm
26
10μm
27
10μm
28
10μm
29
10μm
30
10μm
31
1μm
32
3μm
33
10μm
34
10μm
35
10μm
36
10μm
37
10μm
38
10μm
39
10μm
40
10μm
41
10μm
42
10μm
43
10μm
44
10μm
45
10μm
46
10μm
47
10μm
48
10μm

Appendix 1

The annotated listing of the calcareous dinoflagellate taxa of the investigated material follows the nomenclature of Elbrächter et al., 2008.

Division **Dinoflagellata** (Bütschli, 1885) Fensome et al., 1993

Subdivision **Dinokaryota** Fensome et al., 1993

Class **Dinophyceae** Pascher, 1914

Subclass **Peridiniphycidae** Fensome et al., 1993

Order **Peridiniales** Haeckel, 1894

Suborder **Peridiniineae** Autonym

Family **Thoracosphaeraceae** (Schiller, 1930) Elbrächter et al. 2008

Calciodinellum albatrosianum (Kamptner, 1963) Janofske and Karwath, 2000

Calciodinellum operosum Deflandre, 1947

Calciodinellum levantinum Meier et al., 2002

Calciodinellum limbatum (Deflandre, 1948) Kohring, 1993

Calcigonellum infula (Deflandre, 1948) Keupp, 1984

Caracomia stella (Gilbert and Clark, 1983) Streng et al., 2002

Lebessphaera urania Meier et al., 2002

Leonella granifera (Fütterer, 1977) Janofske and Karwath, 2000

Melodomuncula berlinensis Versteegh, 1993

Pernambugia tuberosa (Janofske and Karwath, 2000)

Pirumella sp. Bison et al., this work

Pirumella parva (Bolli, 1974) Lentin and Williams, 1993

Pirumella rhombica Janofske 1987

Pirumella sicelis Keupp and Kohring, 1993

Praecalcigonellum polymorphum (Keupp 1980)

Praecalcigonellum schizosaeptum Versteegh, 1993

Rhabdothorax spp. (Kamptner, 1937) Kamptner, 1958: includes *Scrippsiella regalis* (Gaardner, 1954) Janofske, 2000 and *Scrippsiella trochoidea* (von Stein, 1983) Loeblich III, 1965

Ruegenia oranensis Keupp and Kohring 1993

Thoracosphaera heimii (Lohmann, 1920) Kamptner, 1944

Appendix 2

lithology	age (~Ma)	cycle	sample no A	sample no B	sample no C	1 No	1 %	2 No	2 %	3 No	3 %	4 No	4 %	5 No	5 %	6 No	6 %	7 No	7 %	8 No	8 %	9 No	9 %	10 No	10 %	11 No	11 %	12 No	12 %	13 No	13 %	14 No	14 %	15 No	15 %	16 No	16 %	17 No	17 %	18 No	18 %	19 No	19 %	20 No	20 %	total no of dinos	diversity
ungenau	~5,23	6	It 17	It 12.017	63	81	59,1	4	2,9	-	-	3	2,2	2	1,5	11	8	-	-	-	-	-	-	-	-	14	10,2	3	2,2	-	-	2	2	-	-	-	-	12	-	2	1,5	-	-	3	2,2	137	1,18
marl	~5,25	4c	It 16	It 12.016	62	22	10,8	-	-	-	-	1	0,5	2	1	98	48	11	5,4	4	2	-	-	-	-	23	11,3	-	-	-	-	-	-	-	-	-	-	17	-	25	12,3	-	-	1	0,5	204	0,92
ungenau	~5,26	4a(2)	It 15	It 12015	61	8	7,4	1	0,9	-	-	2	1,9	-	-	59	54,6	-	-	-	-	-	-	-	-	20	18,5	-	-	1	0,9	-	-	-	-	-	-	4	-	13	12	-	-	-	-	108	1,30
sapropel	~5,27	4a	It 14	It 12.014	60	30	13,6	1	0,5	-	-	3	1,4	2	0,9	107	48,6	13	5,9	4	1,8	-	-	-	-	46	20,9	-	-	-	-	-	-	-	-	-	-	12	-	1	0,5	-	-	1	0,5	220	0,73
marl	~5,27	3c	It 12	It 12.012	59	12	6,7	-	-	-	-	2	1,1	3	1,7	68	38	3	1,7	-	-	-	-	-	-	41	22,9	1	0,6	-	-	-	-	2	1,1	-	-	9	-	37	20,7	1	0,6	-	-	179	1,08
sapropel	~5,29	3a	It 10	It 12.010	58	10	12,2	2	2,4	-	-	3	3,7	-	-	41	50	-	-	-	-	-	-	-	-	12	14,6	-	-	2	2	-	-	-	-	-	-	7	-	5	6,1	-	-	-	-	82	1,77
marl	~5,29	2c	It 8	It 12.008	57	9	3,3	4	1,5	-	-	1	0,4	-	-	228	83,5	3	1,1	-	-	-	-	-	-	14	5,1	-	-	-	-	-	-	1	0,4	-	-	3	-	7	2,6	-	-	3	1,1	273	-0,24
sapropel	~5,31	2a	It 6	It 12.006	56	18	27,3	2	3	-	-	-	-	8	12,1	22	33,3	1	1,5	-	-	-	-	-	-	11	16,7	-	-	-	-	-	-	-	-	-	-	0	-	4	6,1	-	-	-	-	66	2,03
marl?	~5,31	1d	It 5	It 12.005	55	48	25,4	3	1,8	-	-	3	1,6	50	26,5	47	24,9	1	0,5	1	0,5	1	0,5	-	-	12	6,3	-	-	-	-	-	-	-	-	-	-	6	-	11	5,8	-	-	6	3,2	189	1,18
marl	~5,32	1c	It 4	It 12.004	54	6	9,5	2	3,2	-	-	3	4,8	6	9,5	26	41,3	5	7,9	-	-	-	-	-	-	3	4,8	-	-	-	-	-	-	-	-	-	-	9	-	1	1,6	-	-	2	3,2	63	2,33
no sapropel	~5,33	1a	It 2	It 12.002	53	7	9,1	1	1,3	-	-	4	5,2	3	3,9	39	50,6	8	10,4	-	-	-	-	-	-	-	-	-	-	-	-	1	1,3	-	-	3	3,9	7	-	4	5,2	-	-	-	-	77	1,95
marl	5,33	0	M./Pl.	M./Pl.	52	3	0,9	11	3,2	-	-	-	-	-	-	308	90,8	-	-	-	-	-	-	-	-	-	-	-	-	-	-	-	-	-	-	-	-	11	-	7	2,1	-	-	-	-	340	-0,79
CDB	5,97?	50	FT 327	It 19.127	51	-	-	-	-	-	-	-	-	-	-	-	-	-	-	-	-	-	-	-	-	-	-	-	-	-	-	-	-	-	-	-	-	-	-	-	-	-	-	-	-	0	0,00
no sapropel	5,97?	50	FT 324	It 19.124	50	-	-	-	-	-	-	-	-	-	-	-	-	-	-	-	-	-	-	-	-	-	-	-	-	-	-	-	-	-	-	-	-	-	-	-	-	-	-	-	-	0	0,00
marl?	~6,0	48	FT 315	It 19.115	49	-	-	-	-	-	-	-	-	-	-	-	-	-	-	-	-	-	-	-	-	-	-	-	-	-	-	-	-	-	-	-	-	-	-	-	-	-	-	-	-	0	0,00
sapropel	6,015	48	FT 310	It 19.110	48	3	1,3	-	-	-	-	-	-	-	-	1	0,4	-	-	-	-	-	-	-	-	-	-	-	-	-	-	-	-	-	-	-	-	-	-	223	98,2	-	-	-	-	227	-0,72
marl?	~6,05	46	FT 301	It 19.101	47	-	-	-	-	-	-	-	-	-	-	5	100	-	-	-	-	-	-	-	-	-	-	-	-	-	-	-	-	-	-	-	-	-	-	-	-	-	-	-	-	5	3,00
sapropel	6,070	46	FT 294	It 19.094	46	5	71,4	-	-	-	-	-	-	-	-	-	-	-	-	1	14,3	-	-	-	-	-	-	-	-	-	-	-	-	-	-	-	-	1	-	-	-	-	-	-	-	7	3,46
sapropel	6,110	44	FT 282	It 19.082	45	21	42,0	-	-	-	-	-	-	-	-	-	-	-	-	-	-	-	-	-	-	-	-	-	-	-	-	-	-	-	-	-	-	1	-	27	54	-	-	1	2	50	1,55
sapropel	6,164	41	FT 264	It 19.064	44	10	76,9	-	-	-	-	-	-	3	23,1	-	-	-	-	-	-	-	-	-	-	-	-	-	-	-	-	-	-	-	-	-	-	-	-	-	-	-	-	-	-	13	2,58
marl	6,180	40	FT 261	It 19.061	43	2	20,0	-	-	-	-	-	-	2	20	6	60	-	-	-	-	-	-	-	-	-	-	-	-	-	-	-	-	-	-	-	-	-	-	-	-	-	-	-	-	10	3,25
sapropel	6,184	40	FT 257	It 19.057	42	91	75,8	-	-	-	-	-	-	7	5,8	5	4,2	1	0,8	-	-	-	-	-	-	-	-	-	-	-	-	-	-	-	-	-	-	14	-	-	-	-	-	2	1,7	120	0,68
marl	~6,184	39	FT 255	It 19.055	41	1	50,0	-	-	-	-	-	-	-	-	1	50	-	-	-	-	-	-	-	-	-	-	-	-	-	-	-	-	-	-	-	-	-	-	-	-	-	-	-	-	2	4,61
marl	~6,19	39	FT 254	It 19.054	40	8	22,2	-	-	-	-	-	-	-	-	25	69,4	-	-	-	-	-	-	-	-	2	5,6	-	-	-	-	-	-	-	-	-	-	-	-	-	-	-	-	1	2,8	36	1,87
sapropel	6,205	39	FT 250	It 19.050	39	66	93,0	-	-	-	-	-	-	-	-	5	7	-	-	-	-	-	-	-	-	-	-	-	-	-	-	-	-	-	-	-	-	-	-	-	-	-	-	-	-	71	0,60
sapropel	6,205	39	FT 249	It 19.049	38	38	97,4	-	-	-	-	-	-	-	-	1	2,6	-	-	-	-	-	-	-	-	-	-	-	-	-	-	-	-	-	-	-	-	-	-	-	-	-	-	-	-	39	1,06
marl	~6,206	38	FT 248	It 19.048	37	3	100,0	-	-	-	-	-	-	-	-	-	-	-	-	-	-	-	-	-	-	-	-	-	-	-	-	-	-	-	-	-	-	-	-	-	-	-	-	-	-	3	3,51
marl	~6,22	38	FT 247	It 19.047	36	4	80,0	-	-	-	-	-	-	-	-	1	20	-	-	-	-	-	-	-	-	-	-	-	-	-	-	-	-	-	-	-	-	-	-	-	-	-	-	-	-	5	3,50
sapropel	6,226	36	FT 243	It 19.043	35	8	40,0	-	-	-	-	-	-	5	25	7	35	-	-	-	-	-	-	-	-	-	-	-	-	-	-	-	-	-	-	-	-	-	-	-	-	-	-	-	-	20	2,69
sapropel	6,432	28	FT 171	It 18.971	34	138	93,2	-	-	-	-	3	2	-	-	-	-	2	1,4	-	-	-	-	-	-	-	-	-	-	-	-	-	-	-	-	-	-	4	-	1	0,7	-	-	-	-	148	-0,06
sapropel	6,470	26	FT 158	It 18.958	33	2	0,3	-	-	-	-	-	-	-	-	-	-	-	-	-	-	-	-	-	-	-	-	-	-	-	-	-	-	-	-	-	-	1	-	724	99,5	-	-	1	0,1	728	-1,95
marl	6,49	25	FT 156	It 18.956	32	2	66,7	-	-	-	-	-	-	-	-	1	33,3	-	-	-	-	-	-	-	-	-	-	-	-	-	-	-	-	-	-	-	-	-	-	-	-	-	-	-	-	3	4,14
marl	~6,524	23	FT 143	It 18.943	31	1	11,1	-	-	-	-	-	-	-	-	-	-	2	22,2	-	-	-	-	-	-	-	-	-	-	-	-	-	-	-	-	-	-	4	-	1	11,1	-	-	1	11,1	9	3,84
sapropel	6,520	23	FT 139	It 18.939	30	345	88,2	-	-	-	-	1	0,3	-	-	-	-	-	-	-	-	-	-	-	-	-	-	-	-	-	-	-	-	-	-	-	-	10	-	35	9	-	-	-	-	391	-0,93
sapropel	6,585	20	FT 106	It 18.906	29	28	80,0	-	-	-	-	2	5,7	1	2,9	-	-	1	2,9	-	-	-	-	-	-	-	-	-	-	-	-	-	-	-	-	-	-	3	-	-	-	-	-	-	-	35	1,81
marl	~6,62	16	FT 89	It 18.889	28	-	-	-	-	-	-	-	-	-	-	-	-	-	-	-	-	-	-	-	-	-	-	-	-	-	-	-	-	-	-	-	-	-	-	-	-	-	-	-	-	0	0,00
sapropel	6,635	16	FT 78	It 18.878	27	19	67,9	-	-	-	-	-	-	-	-	-	-	-	-	-	-	-	-	-	-	1	3,6	-	-	-	-	-	-	-	-	-	-	5	-	-	-	-	-	3	10,7	28	2,20
marl	~6,658	16	FT 65	It 18.865	26	-	-	-	-	-	-	-	-	-	-	-	-	-	-	-	-	-	-	-	-	-	-	-	-	-	-	-	-	-	-	-	-	-	-	-	-	-	-	-	-	0	0,00
sapropel	6,678	16	FT 60	It 18.860	25	28	66,7	-	-	-	-	1	2,4	-	-	2	4,8	1	2,4	-	-	-	-	-	-	-	-	-	-	-	-	-	-	-	-	-	-	2	-	-	-	-	-	8	19	42	1,92
sapropel	6,771	12	FT 32	It 18.832	24	224	80,6	-	-	-	-	4	1,4	12	4,3	5	1,8	15	5,4	-	-	-	-	-	-	-	-	-	-	-	-	-	-	-	-	-	-	13		2	0,7	-	-	3	1,1	278	-0,19
marl	~6,79	11	FT 29	It 18.829	23	1	6,7	-	-	-	-	-	-	-	-	-	-	2	13,3	-	-	-	-	-	-	9	60	-	-	-	-	-	-	-	-	-	-	1	-	-	-	-	-	2	13,3	15	3,10
marl	~6,78	9/10	FT 23	It 18.823	22	4	7,1	-	-	-	-	-	-	-	-	6	10,7	20	35,7	-	-	-	-	-	-	5	8,9	-	-	-	-	-	-	-	-	-	-	13	-	5	8,9	-	-	3	5,4	56	2,30
sapropel	6,810	9/10	FT 20	It 18.820	21	271	72,1	-	-	-	-	-	-	4	1,1	11	2,9	57	15,2	-	-	-	-	-	-	11	2,9	-	-	-	-	-	-	-	-	-	-	14	-	8	2,1	-	-	-	-	376	-0,34
marl	~7,00?	(0?)	FT 11	It 18.811	20	101	90,2	-	-	-	-	-	-	-	-	4	3,6	-	-	-	-	-	-	-	-	1	0,9	-	-	-	-	-	-	-	-	-	-	2	-	-	-	-	-	4	3,6	112	0,33
marl	~7,02?	(-2?)	FT 5	It 18.805	19	258	94,2	-	-	-	-	2	0,7	-	-	5	1,8	-	-	-	-	-	-	-	-	-	-	-	-	-	-	-	-	-	-	-	-	3	-	3	1,1	-	-	3	1,1	274	-0,69
sapropel	~7,04?	(-2?)	FT 4	It 18.804	18	23	79,3	-	-	-	-	-	-	-	-	-	-	2	6,9	-	-	-	-	-	-	2	6,9	-	-	-	-	-	-	-	-	-	-	-	-	1	3,4	-	-	1	3,4	29	2,02
marl	~7,11	G92	Gib 31	It 14.656	17	133	89,9	-	-	-	-	1	0,7	-	-	-	-	10	6,8	-	-	-	-	-	-	-	-	-	-	-	-	-	-	-	-	2	1,4	1	-	-	-	-	-	1	0,7	148	0,05
marl	~7,11	G92	Gib 32	It 14.655	16	55	74,3	-	-	-	-	-	-	-	-	2	2,7	10	13,5	-	-	-	-	-	-	-	-	-	-	-	-	-	-	-	-	1	1,4	4	-	-	-	-	-	2	2,7	74	1,20
sapropel	~7,13	G92	Gib 34	It 14.653	15	55	82,1	-	-	-	-	1	1,5	-	-	-	-	-	-	-	-	-	-	-	-	4	6	-	-	-	-	-	-	-	-	2	3	1	-	-	-	-	-	4	6,0	67	1,13
sapropel	7,130	G92	Gib 35	It 14.652	14	106	80,3	-	-	1	0,3	4	3	2	1,5	6	4,5	9	6,8	-	-	-	-	-	-	1	0,8	-	-	-	-	-	-	-	-	-	-	1	-	-	-	-	-	2	1,5	132	0,57
sapropel	7,150	G91	Gib 40	It 14.647	13	300	91,5	-	-	-	-	1	0,3	5	1,5	-	-	10	3	-	-	-	-	-	-	4	1,2	-	-	-	-	-	-	-	-	-	-	6	-	-	-	-	-	2	0,6	328	-0,76
marl	~7,15	G90	Gib 42	It 14.645	12	227	88,7	-	-	-	-	1	0,4	1	0,4	-	-	8	3,1	1	0,4	-	-	-	-	2	0,8	-	-	-	-	-	-	-	-	-	-	7	-	-	-	-	-	9	3,5	256	-0,41
sapropel	7,170	G90	Gib 44	It 14.643	11	60	52,6	-	-	-	-	1	0,9	4	3,5	7	6,1	1	0,9	-	-	-	-	-	-	24	21,1	-	-	-	-	-	-	-	-	-	-	10	-	-	-	-	-	7	6,1	114	1,29
marl	~7,17	G89	Gib 47	It 14.640	10	241	86,4	-	-	-	-	-	-	11	3,9	3	1,1	1	0,4	-	-	-	-	-	-	8	2,9	-	-	-	-	-	-	-	-	-	-	11	-	-	-	-	-	4	1,4	279	-0,41
sapropel	7,190	G89	Gib 50	It 14.637	9	264	93,0	-	-	-	-	-	-	1	0,4	4	1,4	1	0,4	2	0,7	-	-	-	-	5	1,8	-	-	-	-	-	-	-	-	-	-	1	-	-	-	-	-	6	2,1	284	-0,67
marl	~7,19	G88	Gib 52	It 14.635	8	151	95,6	-	-	-	-	-	-	-	-	1	0,6	-	-	-	-	-	-	-	-	-	-	-	-	-	-	-	-	-	-	-	-	-	-	-	-	-	-	6	3,8	158	-0,26
sapropel	7,212	G88	Gib 55	It 14.632	7	57	91,9	-	-	-	-	-	-	-	-	1	1,6	1	1,6	-	-	-	-	-	-	3	4,8	-	-	-	-	-	-	-	-	-	-	-	-	-	-	-	-	-	-	62	0,84
marl	~7,22	G87	Gib 57	It 14.630	6	324	87,8	-	-	-	-	-	-	1	0,3	-	-	12	3,3	2	0,5	-	-	-	-	8	2,2	-	-	-	-	-	-	-	-	3	0,8	1	-	-	-	-	-	18	4,9	369	-0,75
sapropel	7,242	G87	Gib 59	It 14.628	5	127	67,2	-	-	1	0,1	-	-	-	-	-	-	22	11,6	-	-	-	-	-	-	25	13,2	-	-	-	-	-	-	-	-	4	2,1	-	-	2	1,1	-	-	8	4,2	189	0,44
sapropel	~7,51	G86	Gib 60	It 14.627	4	1	33,3	-	-	-	-	-	-	-	-	-	-	-	-	-	-	-	-	-	-	-	-	-	-	-	-	-	-	-	-	2	66,7	-	-	-	-	-	-	-	-	3	4,14
marl	~7,53	G85	Gib 61	It 14.626	3	37	80,4	-	-	-	-	1	2,2	-	-	-	-	2	4,3	-	-	-	-	-	-	-	-	-	-	-	-	-	-	-	-	5	10,9	-	-	1	2,2	-	-	-	-	46	1,50
marl	~7,53	G85	Gib 62	It 14.625	2	54	73,0	-	-	-	-	-	-	-	-	1	1,4	-	-	-	-	-	-	2	2,7	-	-	-	-	-	-	-	-	-	-	10	13,5	2	-	-	-	-	-	-	6,8	69	1,24
marl	~7,53	G85	Gib 63	It 14.624	1	27	64,3	-	-	-	-	-	-	-	-	-	-	12	28,6	-	-	-	-	-	-	-	-	-	-	-	-	-	-	-	-	3	7,1	-	-	-	-	-	-	-	-	42	1,70
			sum per species:			4178		31		2		48		135		1170		247		15		1		2		311		4		3		3		3		35		233		1150		1		118		sum: 7690	

Table 1: Count data and calculated diversity data for all analysed samples.
1 = *Calciodinellum albatrosianum*, 2 = *Calciodinellum levantinum,* 3 = *Calciodinellum limbatum,* 4 = *Calciodinellum operosum,* 5 = *Caracomia stella,* 6 = *Leonella granifera,* 7 = *Lebessphaera urania,* 8 = *Melodomuncula berlinensis*, 9 = *Calcigonellum infula,* 10 = *Ruegenia oranensis,* 11 = *Pirumella parva,* 12= *Pirumella polymorphum*, 13 = *Pirumella rhombica,* 14 = *Praecalcigonellum schizosaeptum*, 15 = *Pirumella sicelis,* 16 = *Pirumella tuberosa*, 17 = *Rhabdothorax* spp., 18 = *Thoracosphaera heimii,* 19 = *Pirumella sp.,* 20 = *not identified.* A = field number, B = original sample number, C = sample number (this work).

References

Arnaboldi, M., Meyers, P.A., 2006. Patterns of organic carbon and nitrogen isotopic compositions of latest Pliocene sapropels from six locations across the Mediterranean Sea. *Palaeogeography, Palaeoclimatology, Palaeoecology* **235**: 149-167.

Bellanca, A., Caruso, A., Ferruzza, G., Neri, R., Rouchy, J.-M., Sprovieri, M., Blanc-Valleronc, M.-M., 2001. Transition from marine to hypersaline conditions in the Messinian Tripoli Formation from the marginal areas of the central Sicilian Basin. *Sedimentary Geology*, **140**(1-2): 87-105

Bison, K.-M., Versteegh, G.J.M., Hilgen, F.J., Willems, H., 2007. Calcareous dinoflagellate turnover in relation to the Messinian salinity crisis in the eastern Mediterranean Pissouri Basin, Cyprus. *Journal of Micropalaeontology*, **26**: 103-116.

Bison, K.-M., Versteegh, G.J.M., Orszag-Sperber, F., Rouchy, J.-M., Willems, H., 2009. Palaeoenvironmental changes of the early Pliocene (Zanclean) in the eastern Mediterranean Pissouri Basin (Cyprus) evidenced from calcareous dinoflagellate cyst assemblages. *Marine Micropaleontology*, **73**: 49-56.

Blanc, P.-L., 2000. Of sills and straits: a quantitative assessment of the Messinian Salinity Crisis. *Deep-Sea Research*, **47**, 1429–1460.

Bouchet, P., Taviani, M. 1992. The Mediterranean deep-sea fauna: pseudopopulations of Atlantic species? *Deep-sea Research*, **39**: 169-184.

Bouillon, J., Medel, M.D., Pagos, F., Gili, J.-M., Boero, F., Gravili, C. 2004. Fauna of the Mediterranean Hydrozoa. *Scientia Marina* (Barcelona), *Consejo Superior de Investigaciones Cientificas*, Institut de Ciènces del Marina, Barcelona, **68**(2): 5-449.

Brand, L.E., Sunda. W.G., Guillard, R.R.L., 1983. Limitation of marine phytoplankton reproductive rates by zinc, manganese and iron. *Limnology and Oceanography*, **28**:1182-1198

Butler, R.W.H., McClelland, E., Jones, R.E. 1999. Calibrating the duration and timing of the Messinian salinity crisis in the Mediterranean: linked tectonoclimatic signals in thrust-top basins in Sicily. *Journal of the Geological Society, London*, **156**: 827-835.

Cita, M.B., Gartner, S., 1973. The Stratotype Zanclean: foraminiferal and nannofossil biostratigraphy. *Rivista Italiana di Paleontologia i Stratigrafia*, **79**: 503– 558.

Cramp, A., Collins, M.B., West, R., 1988. Late Pleistocene–Holocene sedimentation in the NW Aegean Sea: a palaeoclimatic palaeoceanographic reconstruction. *Palaeogeography, Palaeoclimatology, Palaeoecology*, **68**: 61–77.

Dale, B., 1992b. Thoracosphaerids: Pelagic Fluxes. In: S. Honjo (Editor), Dinoflagellate Contributions to the Deep Sea. *Ocean Biocoenosis Series*, **5**. *Woods Hole Oceanographic Instituion*, Massachusetts: 33-44.

De Lange, G. J., van Santvoort, P. J. M., Langereis C., Thomson, J., Corselli, C., Michard, A. Rossignol-Strick, M., Paterne, M., Anastasakis, G. 1999. Palaeo-environmental variations in eastern Mediterranean sediments: A multidisciplinary approach in a prehistoric setting. *Progress in Oceanography*, **44**: 369–386.

De Lange, G.J., Thomson, J., Reitz, A., Slomp, C.P., Speranza Principato, M., Erba, E. , Corselli, C., 2008. Synchronous basin-wide formation and redox-controlled preservation of a Mediterranean sapropel. *Nature Geoscience*, **1**: 606-610.

Decima, A., Wezel, F. C., 1973. Late Miocene evaporites of the Central Sicilian Basin, Sicily.- In: Ryan, W.B.F., Hsü, K.J. et al.(Eds.). *Initial Reports Deep-Sea Drilling Project*, **13**(2): 1234-1240,

Elbrächter, M., Gottschling, M., Hildebrand-Habel, T., Keupp, H., Kohring, R., Lewis, J., Meier, S.K.J., Montresor, M., Streng, M., Versteegh, G.J.M., Willems, H., Zonneveld, K.A.F., 2008. Establishing an Agenda for Calcareous Dinoflagellate Research (Thoracosphaeraceae, Dinophyceae) including a nomenclatural synopsis of generic names. *Taxon*, **57**: 1289-1303.

Esper, O., Zonneveld, K. A. F., Höll, C., Karwath, B., Schneider, R., Vink, A., Weise-Ihlo, I., Willems, H., 2000. Reconstruction of palaeoceanographic conditions in the South Atlantic Ocean at the last two Terminations based on calcareous dinoflagellates. *International Journal of Earth Sciences*, **88**(4): 680-693.

Fauquette, S., Guiot, J., Suc, J.P., 1998. A method for climatic reconstruction of the Mediterranean Pliocene using pollen data. *Palaeogeography, Palaeoclimatology, Palaeoecology*, **144**: 183-201.

Fauquette, S., Suc, J., Bertini, a, Popescu, S., Warny, S., Bachiritaoufiq, N., Perezvilla, M., et al., 2006. How much did climate force the Messinian salinity crisis? Quantified climatic conditions from pollen records in the Mediterranean region. *Palaeogeography, Palaeoclimatology, Palaeoecology*, **238**: 281-301.

Fontugne, M.R., Calvert, S.E., 1992. Late Pleistocene variability of the carbon isotopic composition of organic matter in the eastern Mediterranean Sea: monitor of changes in carbon sources and atmospheric CO_2 concentrations. *Paleoceanography*, **71**: 1–20.

Frydas, D., Keupp, H., 1992: Kieseliges und kalkiges Phytoplankton aus dem Neogen von NW- und W-Kreta/Griechenland. *Berliner geowissenschaftliche Abhandlungen*, Berlin, **E**(3): 97-111;

Fütterer, D., 1977. Distribution of calcareous dinoflagellates in Cenozoic sediments of Site 366, Eastern North Atlantic. *Initial Reports of the Deep Sea Drilling Project*, **41**: 533-541.

Fütterer, D., 1984. Pithonelloid dinoflagellates from the Upper Cretaceous and Cenozoic of the southeastern Atlantic ocean, deep sea drilling project Leg 74. *Initial Reports of the Deep Sea Drilling Project*, **74**: 533–541.

Gaines, G., Elbrächter, M., 1987. Heterotrophic nutrition In: Taylor, F.J.R. (Ed.). The biology of dinoflagellates. *Blackwell Scientific Publications*, **36**: 224-268.

Geider, R.J., Roche, J.L., 1994. The role of iron in phytoplankton photosynthesis, and the potential for iron-limitation of primary productivity in the sea. *Limnology*: 275-301.

Geider, R.J., 1999. Complex lessons of iron uptake. *Nature*, **400**: 115-116.

Gersonde, R. and Schrader, H., 1984. Marine planktic diatom correlation of lower Messinian deposits in the Western Mediterranean. *Marine Micropalaeontology*. **9**: 93-110.

Griffin, D.L., 2002. Aridity and humidity: two aspects of the late Miocene climate of North Africa and the Mediterranean. *Palaeogeography, Palaeoclimatology Palaeoecology*, **182**: 65–91.

Hildebrand-Habel, T., Willems, H., Versteegh, G.J.M., 1999. Variations in calcareous dinoflagellate associations from the Maastrichtian to Middle Eocene of the western South Atlantic Ocean (Sao Paulo Plateau, DSDP Leg 39, Site 356). *Review of Palaeobotany and Palynology*, **106**(1-2): 57-87.

Hildebrand-Habel, T., Willems, H., 2000. Distribution of calcareous dinoflagellates from the Maastrichtian to early Miocene of DSDP Site 357 (Rio Grande Rise, western South Atlantic Ocean). *International Journal of Earth Sciences*, **88**: 694-707.

Hildebrand-Habel, T., Streng, M., 2003. Calcareous dinoflagellate associations and Maastrichtian–Tertiary climatic change in a high latitude core (ODP Hole 689B, Maud Rise, Weddell Sea). *Palaeogeography, Palaeoclimatology, Palaeoecology*, **197**: 293–321.

Hilgen, F.J., 1987, Sedimentary rhythms and high-resolution chronostratigraphic correlations in the Mediterranean Pliocene: *Newsletters in Stratigraphy*, **17**: 109-127.

Hilgen, F.J., 1991. Astronomical calibration of Gauss to Matuyama sapropels in the Mediterranean and implication for the geomagnetic polarity time scale. *Earth and Planetary Science Letter*, **104**: 226-244.

Hilgen, F.J., Krijgsman, W., 1999. Cyclostratigraphy and astrochronology of the Tripoli diatomite formation (pre-evaporite Messinian Salinity Crisis, Italy). *Terra Nova*, **11**: 16–22.

Hilgen, F.J., Krijgsman, W., Langereis, C.G., Lourens, L.J., Santarelli, A., Zachariasse, W.J., 1995. Extending the astronomical (polarity) time scale into the Miocene. Earth Planetary Science Letters, **136**: 495–510.

Hilgen, F.J., Krijgsman, W., Raffi, I., Turco, E., Zachariasse, W.J., 2000. Integrated stratigraphy and astronomical calibration of the Serravallian/Tortonian boundary section at Monte Gibliscemi (Sicily, Italy). *Marine Micropaleontology*, **38**: 181–211.

Hilgen, F.J., Abdul Aziz, H., Bice, D., Iaccarino, S., Krijgsman, W., Kuiper, K.F., Montanari, A., Raffi, I., Turco, E., Zachariasse, W.J., 2005. The Global Boundary Stratotype Section and Point (GSSP) of the Tortonian Stage (Upper Miocene) at Monte dei Corvi. *Episodes*, **28**: 6–17

Hilgen, F., Kuiper, K., Krijgsman, W., Snel, E., Van der Laan, E., (2007). Astronomical tuning as the basis for high resolution chronostratigraphy: the intricate history of the Messinian Salinity Crisis. *Stratigraphy*, **4**: 231–238.

Höll, C., Zonneveld, K.A.F., Willems, H. 1998. On the ecology of calcareous dinoflagellates: The Quaternary Eastern Equatorial Atlantic. *Marine Micropaleontology*, **33**(1-2): 1-25.

Hüsing, S.K., Kuiper, K.F., Link, W., Hilgen, F.J., Krijgsman, W., 2009. The upper Tortonian–lower Messinian at Monte dei Corvi (Northern Apennines, Italy): Completing a Mediterranean reference section for the Tortonian Stage. *Earth and Planetary Science Letters*, **282**: 140–157.

Inouye, I., Pienaar, R.N., 1983. Observations on the life cycle and microanatomy of Thoracosphaera heimii (Dinophyceae) with special reference to its systematic position. *South African Journal of Botany*, **2**: 63–75.

Janofske, D. 1996. Ultrastructure types in recent "Calcispheres". *Bulletin de l'Institut océanographique*, **14**(4): 295-303.

Janofske, D., 2000. *Scrippsiella trochoidea* and *Scrippsiella regalis*, nov. comb. (Peridiniales, Dinophyceae): a comparison. *Journal of Phycology*, **36**: 178–189.

Janofske, D., Karwath, B., 2000. Oceanic calcareous dinoflagellates of the equatorial Atlantic Ocean: Cyst-theca relationship, taxonomy and aspects on ecology. *In*: Karwath, 2000, Ecological Studies on Living and Fossil Calcareous Dinoflagellates of the Equatorial and Tropical Atlantic Ocean. *Berichte Fachbereich Geowissenschaften*, Universität Bremen, **152**:94-136.

Jickells, T.D., An, Z.S., Andersen, K.K., Baker, A.R., Bergametti, G., Brooks, N., Cao, J.J., Boyd, P.W., Duce, R.A., Hunter, K.A., Kawahata, H., Kubilay, N., La Roche, J., Liss, P.S., Mahowald, N., Prospero, J.M., Ridgwell, A.J., Tegen, I., Torres, R., 2005. Global iron connections between desert dust, ocean biogeochemistry, and climate. *Science*, **308**: 67–71.

Johnson, K.S., Chavez, F.P., Friederich, G.E., 1999. Continental-shelf sediment as a primary source of iron for coastal phytoplankton. *Nature*: 697-700.

Karwath, B., Janofske, D., Tietjen, F., Willems, H., 2000a. Temperature effects on growth and cell size in the marine calcareous dinoflagellate *Thoracosphaera heimii*. *Marine Micropaleontology*, **39**(1-4): 43-51.

Karwath, B., Janofske, D., Willems, H., 2000b. Spatial distribution of the calcareous dinoflagellate *Thoracosphaera heimii* in the upper water column of the tropical and equatorial Atlantic. *International Journal of Earth Sciences*, **88**: 668-679.

Keupp, H., 1987. Die kalkigen Dinoflagellatenzysten des Mittelalb bis Untercenoman von Escalles/Boulonnais (N-Frankreich). *Facies*, **16**: 37- 88.

Keupp, H., Kohring, R., 1993. Kalkige Dinoflagellatenzysten aus dem Obermiozän von El Medhi (Algerien). *Berliner Geowisswissenschaftliche Abhandlungen*, **9**: 25-43.

Keupp, H., Bellas, S.M., Frydas, D., Kohring, R., 1994. Aghia Irini, ein Neogenprofil auf der Halbinsel Gramvoússa/NW-Kreta. *Berliner Geowissenschaftliche Abhandlungen*, **13**(E): 469-481.

Keupp, H., Kohring, R., 1999. Kalkiger Dinoflagellatenzysten aus dem Obermiozän (NN 11) W von Rethymnon (Kreta). *Berliner Geowissenschaftliche Abhandlungen*, **30**(E): 33-53.

Keupp, H., Versteegh, G., 1989. Ein neues Konzept für kalkige Dinoflagellaten-zysten der Subfamilie Orthopithonelloideae Keupp 1987. *Berliner Geowissenschaftliche Abhandlungen*, **106**(A): 207-219.

Kohring, R., 1993a. Kalkdinoflagellaten aus dem Mittel- und Obereozän von Jütland (Dänemark) und dem Pariser Becken (Frankreich) im Vergleich mit anderen Tertiär-Vorkommen. *Berliner Geowissenschaftliche Abhandlungen*, **6**(E): 1-164.

Kohring, R., 1993b. Kalkdinoflagellaten-Zysten aus dem unteren Pliozän von E-Sizilien. *Berliner Geowissenschaftliche Abhandlungen*, **9**(E): 15-23.

Kohring, R., 1997. Calcareous dinoflagellate cysts from the Blue Clay formation (Serravalian, Late Miocene) of the Maltese Islands. *Neues Jahrbuch, Geologisch-Paläontologische Mitteilungen*, **3**: 151-164.

Kohring, R., Gottschling, M., Keupp, H., 2005. Examples for character traits and palaeoecological significance of calcareous dinoflagellates. *Paläontologische Zeitschrift*, **79**(1): 79-91.

Kouwenhoven, T.J., Hilgen, F.J., van der Zwaan, G.J., 2003. Late Tortonian–early Messinian stepwise disruption of the Mediterranean–Atlantic connections: constraints from benthic foraminiferal and geochemical data. *Palaeogeography, Palaeoclimatology, Palaeoecology*, **198**: 303–319.

Kouwenhoven, T.J., Van der Zwaan, G.J., 2006. A reconstruction of late Miocene Mediterranean circulation patterns using benthic foraminifera. *Palaeogeography, Palaeoclimatology, Palaeoecology,* **238**: 373–385.

Kouwenhoven, T. J., Morigi, C., Negri, A., Giunta, S., Krijgsman, W., Rouchy, J.-M., 2006. Paleoenvironmental evolution of the eastern Mediterranean during the Messinian: Constraints from integrated microfossil data of the Pissouri Basin (Cyprus). *Marine Micropaleontology*, **60**(1): 17-44.

Kovar-Eder, J., Kvaček, Z., Martinetto, E., Roiron, P., 2006. Late Miocene to Early Pliocene vegetation of southern Europe. In: Agustí, J., Oms, J., Meulenkamp, J.E., (Eds.). Late Miocene to Early Pliocene environment and climate change in the Mediterranean area. *Palaeogeography, Palaeoclimatology, Palaeoecology*, **238**: 321–339.

Krijgsman, W., Hilgen, F.J., Langereis, C.G., Santarelli, A. and Zachariasse, W.J., 1995. Late Miocene magnetostratigraphy, biostratigraphy and cyclostratigraphy in the Mediterranean. *Earth and Planetary Science Letters*, **136**: 475-494.

Krijgsman, W., Langereis, C.G., Zachariasse, W.J., Boccaletti, M., Moratti, G., Gelati, R., Iaccarinof, S., Papani, G., Villa, G., 1999. Late Neogene evolution of the Taza–Guercif Basin (Rifian Corridor,Morocco) and implications for the Messinian salinity crisis. *Marine Geology*, **153**:147–160.

Krijgsman, W., Blanc-Valleron, M.-M., Flecker, R., Hilgen, F. J., Kouwenhoven, T. J., Merle, D., Orszag-Sperber, F., Rouchy, J.-M., 2002. The onset of the Messinian salinity crisis in the eastern Mediterranean (Pissouri Basin, Cyprus). *Earth Planetary Science Letters*, **194**: 299– 300.

Krijgsman, W., Garcés, M., 2004. Paleomagnetic constraints on the geodynamic evolution of the Gibraltar Arc. *Terra Nova*, **16**: 281–287.

Krijgsman W., Meijer, P.T., 2008. Depositional environments of the Mediterranean "Lower Evaporites" of the Messinian Salinity Crisis: constraints from quantitative analyses. *Marine Geology*, **253**(3–4): 73–81.

Lam, P.J., Bishop, J.K.B., 2008. The continental margin is a key source of iron to the HNLC North Pacific Ocean. *Geophysical Research Letters*, **35**: L07608, doi:10.1029/2008GL033294.

Langereis, C.G., Hilgen, F.J., 1991. The Rossello composite: a Mediterranean and global reference section for the Early to early Late Pliocene. *Earth and Planetary Science Letters*, **104**(2-4): 211-225.

Logan, A., Bianchi, C.N., Morri, C., Zibrowius, H., 2004. The present-day Mediterranean brachiopod fauna: diversity, life habits, biogeography and paleobiogeography. *Scientia Marina*, **68**(1): 167-170.

Londeix, L., Benzakour, M., Suc, J.-P., Turon, J.-L., 2007. Messinian palaeoenvironments and hydrology in Sicily (Italy): The dinoflagellate cyst record Pale´oenvironnements et hydrologie du Messinien de Sicile (Italie) d'apre`s les kystes de dinoflagelle´s. *Geobios*, **40**: 233–250.

Martin J.H., Coale K.H., Johnson K.S., Fitzwater S.E., Gordon R.M., Tanner S.J., Hunter C.N., Elrod V.A., Nowiicki J.L., Coley T.L., Barber R.T., Lindley S., Watson A.J., Van Scoy K., C.S. Law, M.I. Liddicoat, R. Ling, T. Stanton, J. Stockel, C. Collins, A. Anderson, R. Bidigare, M. Ondrusek, M. Latasa, F.J. Millero, K. Lee, W. Yao, J.Z. Zhang, G. Friederich, C. Sakamoto, F. Chavez, K. Buck, Z. Kolber, R. Greene, P. Falkowski, S.W. Chisholm, F. Hoge, R. Swift, J. Yungel, S. Turner, S., Nightingale, P., Hatton, A., Liss, P., Tindale, N.W., 1994. Testing the iron hypothesis in ecosystems of the equatorial Pacific Ocean. *Nature*, **371**: 123–129.

McKenzie, J.A., Jenkyns, H., Bennett, G., 1979. Stable isotope study of the cyclic diatomite—clay-stones from the Tripoli Formation, Sicily: a prelude to the Messinian salinity crisis, *Palaeogeography, Palaeoclimatology, Palaeoecology,* **29**: 125–141.

Meier, K.J.S., Janofske, D, Willems, H. 2002. New calcareous dinoflagellates (Calciodinelloideae) from the Mediterranean Sea. *Journal of Phycology*, **38**: 602-615.

Meier, K.J.S., Willems, H., 2003. Calcareous dinoflagellate cysts from surface sediments of the Mediterranean Sea: distribution patterns and influence of main environmental gradients. *Marine Micropaleontology*, **48**: 321-354.

Meier, K.J.S., Zonneveld, K.A.F., Kasten, S., Willems, H., 2004a. Different nutrient sources forcing increased productivity during eastern Mediterranean S1 sapropel formation as reflected by calcareous dinoflagellate cysts. *Paleoceanography*, **19**: 1–12.

Meier, K.J.S., Höll, C., Willems, H., 2004b. Effect of temperature on culture growth and cyst production in the calcareous dinoflagellates *Calciodinellum albatrosianum*, *Leonella granifera* and *Pernambugia tuberosa*. In: M. Triantaphyllou (Ed), Advances in the biology, ecology and taphonomy of extant calcareous nannoplankton. *Micropaleontology*, **50**(1): 93-106.

Meijer, P.T., Krijgsman, W., 2005. A quantitative analysis of the desiccation and re-filling of the Mediterranean during the Messinian Salinity Crisis. *Earth and Planetary Science Letters*, **240**(2): 510-520.

Meyers, P.A., 2006. Paleoceanographic and paleoclimatic similarities between Mediterranean sapropels and Cretaceous black shales. *Palaeogeography, Palaeoclimatology, Palaeoecology*, **235**(1-3): 305 - 320.

Meulenkamp, J., Dermitzakis, M., Georgiadou-Dikeoulia, E., Jonkers, H.A., Böger, H., 1979. Field guideto the Neogene of Crete. *Publications of the Department of Geology and Paleontology*, University of Athens, **A**: 1-15.

Montresor, M., Zingone, A. and Sarno, D., 1998. Dinoflagellate cyst production at a coastal Mediterranean site. *Journal of Plankton Research*, **20**(12): 2291-2312.

Morel, F.M.M., Rueter, J.G., Price, N.M, 1991. Iron nutrition of phytoplankton and its possible importance in the ecolo- gy of ocean regions with high nutrient and low biomass. *Oceanography*, **4**: 56-61.

Passier, H.F., Bosch, H.-J., Nijenhuis, I.A., Lourens, L.J., Bottcher, M.E., Leenders, A., Sinninghe Damsté, J.S., De Lange, G.J., De Leeuw, J.W., 1999. Sulphidic Mediterranean surface waters during Pliocene sapropel formation. *Nature*, **397**: 146–149.

Passier, H.F., Middelburg, J.J., De Lange, G.J. and Bottcher, M.E., 1999. Modes of sapropel formation in the eastern Mediterranean: some constraints based on pyrite properties. *Marine Geology*, **153**: 199–219.

Richter, D., Vink, A., Zonneveld, K.A.F., Kuhlmann, H., Willems, H., 2007. Calcareous dinoflagellate cyst distributions in surface sediments from upwelling areas off NW Africa, and their relationships with environmental parameters of the upper water column. *Marine Micropaleontology*, **63**(3–4): 201–228.

Richter, D., 2009. Characteristics of calcareous dinoflagellate cyst assemblages in a major upwelling region (NW Africa) Spatial distribution , fluxes and ecology. *Dissertation zur Erlangung des Doktorgrades der Naturwissenschaften,* Fachbereich Geowissenschaften, Universität Bremen: 140pp.

Rohling, E.J., Hilgen, F.J., 1991. The eastern Mediterranean climate at times of sapropel formation: a review. *Geologie en Mijnbouw*: 253–264.

Rohling, E.J., 1994. Review and new aspects concerning the formation of eastern Mediterranean sapropels. *Marine Geology*, **122**: 1–28.

Rossignol-Strick, M., Nesteroff, W., Olive, P., and Vergnaud-Grazzini, C., 1982. After the deluge: Mediterranean stagnation and sapropel formation. *Nature*, **295**:105–110.

Rossignol-Strick , M., 1983. African monsoons, an immediate climate response to orbital insolation. *Nature*, **304**: 46–49.

Rossignol-Strick, M. 1985. Mediterranean Quaternary sapropels, an immediate response of the African monsoon to variation of insolation. *Palaeogeography, Palaeoclimatology, Palaeoecology*, **49**: 237-263.

Rouchy, J.M., Caruso, A., 2006. The Messinian Salinity Crisis in the Mediterranean basin: a reassessment of the data and an integrated scenario. *Sedimentary Geology*, **188–189**: 35–67.

Ruggieri, G., Sprovieri, R., 1976. Messinian salinity crisis and its paleogeographical implications. *Palaeogeography, Palaeoclimatology, Palaeoecology*, **20**: 13–21.

Ryan, W.B.F., et al., 1973. *Initial Reports of the Deep Sea Drilling Project*, **13**: 403–464.

Sarthou, G., Baker, A.R., Blain, S., Achterberg, E.P., Boye, M., Bowie, A.R., Croot, P., Laan, P., Baar, H.J.W.d., Jickells, T.D., Worsfold, P.J., 2003. Atmospheric iron deposition and sea-surface dissolved iron concentrations in the eastern Atlantic Ocean. *Deep Sea Research*, **50**(I10-11): 1339-1352.

Shannon, C.E., Weaver, W., 1949. The Mathematical Theory of Communication. *University of Illinois Press*, Urbana Champaign.

Shi, Z., Krom, M.D., Bonneville, S., 2009. Formation of Iron Nanoparticles and Increase in Iron Reactivity in Mineral Dust during Simulated Cloud Processing. *Environmental Science & Technology*, **43**(17): 6592–6596.

Siggelkow, D., Vink, A., Willems, H., 2002. Calcareous dinoflagellate cyst production, vertical transport and preservation off Cape Blanc during 1990: a sediment trap study. *Journal of Nannoplankton Research*, **24**(2): 160.

Spezzaferri, S., Cita, M.-B., McKenzie, J. A., 1998. The Miocene/Pliocene boundary in the eastern Mediterranean: Results from sites 967 and 969. *Proceedings of the Ocean Drilling Program, Scientific Results,* **160**: 9-28.

Sprovieri, R., Di Stefano, E., Sprovieri, M., 1996a. High resolution chronology for late Miocene Mediterranean stratigraphic events. *Rivista Italiana di Paleontologia e Stratigrafia*, **102**: 77–104.

Sprovieri, R., Di Stefano, E., Caruso, A., Bonomo, S., 1996b. High resolution stratigraphy in the Messinian Tripoli Formation in Sicily. *Paleopelagos*, **6**: 415–435.

Sprovieri, R., Thunell, R.C., 1997. Pliocene sapropels in the northern Adriatic area : chronology and paleoenvironmental significance. *Science*, **135**: 1-25.

Streng, M., Hildebrand-Habel, T. and Willems, H. 2002. Revision of the genera *Sphaerodinella* Keupp and Versteegh, 1989 and *Orthopithonella* Keupp *in* Keupp and Mutterlose, 1984 (Calciodinelloideae, calcareous dinoflagellate cysts). *Journal of Paleontology*, **76**: 397-407.

Streng, M., Hildebrand-Habel, T., Willems, H., 2004. Long-term evolution of calcareous dinoflagellate associations since the Late Cretaceous: comparison of a high- and a low-latitude core from the Indian Ocean. *Journal of Nannoplankton Research*, **26**: 13-45.

Suc, J.-P., Violanti, D., Londeix, L., Poumot, C., Robert, C., s Clauzon, G., s Gautier, F., Turon, J.-L., Ferrier, J. Chikhi, H., Cambon, G., 1995. Evolution of the Messinian Mediterranean environments: the Tripoli Formation at Capodarso (Sicily, Italy). *Review of Palaeobotany and Palynology*, **87**(1): 51-79.

Sundareshwar, P.V., Morris, J.T., Koepfler, E.K., Fornwalt, B., 2003. Phosphorus limitation of coastal ecosystem processes. *Science*, **299**: 563-565.

Sundareshwar, P.V., Murtugudde, R., Srinivasan, G., Singh, S., Ramesh, K.J., Ramesh, R., Verma, S.B., Agarwal, D., Baldocchi, D., Baru, C.K., Baruah, K.K., Chowdhury, G.R., Dadhwal, V.K., Dutt, C.B.S., Fuentes, J., Gupta, P.K., Hargrove, W.W., Howard, M., Jha, C.S., Lal, S, Michener, W.K., Mitra, A.P., Morris, J.T., Myneni, R.R., Naja, M., Nemani, R., Purvaja, R., Raha, S., Santhana Vanan, S.K., Sharma, M., Subramaniam, A., Sukumar, R., Twilley, R.R., Zimmerman, P.R., 2007. Environmental Monitoring Network for India. *Science*, **5822**(316): 204–205.

Tangen, K., Brand, L.E., Blackwelder, P.L. and Guillard, R.R.L., 1982. Thoracosphaera heimii (Lohmann) Kamptner is a dinophyte: Observations on its morphology and life cycle. *Marine Micropaleontology*, **7**: 193-212.

Tanimura, Y., Shimada, C., 2004. Calcareous dinoflagellates from a northwestern Pacific sediment trap and their paleoceanographic implications. *Micropaleontology*, **50**: 343-356.

Thunell, R.C., Williams, D.F., Belyea, P.R., 1984. Anoxic events in the Mediterranean Sea in relation to the evolution of late Neogene climates. *Marine Geology*, **59**: 105–134.

Van Assen, E., Kuiper, K.F., Barhoun, N., Krijgsman, W., Sierro, F.J., 2006. Messinian astrochronology of the Melilla Basin: stepwise restriction of the Mediterranean–Atlantic connection through Morocco. *Palaeogeography Palaeoclimatology Palaeoecology*, **238**: 15–31.

Van der Zwaan, G.J., Gudjonsson, L., 1986. The Middle Miocene–Pliocene stable isotope stratigraphy and paleoceanography of the Mediterranean. *Marine* Micropaleontology, **10**: 71–90.

Versteegh, G.J.M., 1993. New Pliocene and Pleistocene calcareous dinoflagellate cysts from southern Italy and Crete. *Review of Palaeobotany and Palynology*, **78**: 353-380.

Vink, A., Zonneveld, K.A.F., Willems, H., 2000. Distributions of calcareous dinoflagellates in surface sediments of the western equatorial Atlantic, and their potential use in palaeoceanography. *Marine Micropaleontology*, **38**: 149-180.

Vink, A., Rühlemann, C., Zonneveld, K.A.F., Mulitza, S., Hüls, M., Willems, H., 2001. Shifts in the position of the North Equatorial Current and rapid productivity changes in the western Tropical Atlantic during the last glacial. *Paleoceanography*, **16**(5): 479-490.

Vink, A., Brune, A., Höll, C., Zonneveld, K.A.F., Willems, H., 2002. On the response of calcareous dinoflagellates to oligotrophy and stratification of the upper water column in the equatorial Atlantic Ocean. *Palaeogeography Palaeoclimatology Palaeoecology*, **178**: 53-66.

Vink, A., Baumann, K.-H., Boeckel, B., Esper, O., Kinkel, H., Volbers, A., Willems, H., Zonneveld, K.A.F., 2003. Coccolithophorid and dinoflagellate synecology in the South and Equatorial Atlantic: Improving the palaeoecological significance of phytoplanktonic microfossils, *In*: Wefer, G; Mulitza, S.,

Rathmeyer, V. (Eds.). The South Atlantic in the Late Quaternary: Reconstruction of Material Budgets and Current Systems. *Springer Verlag*, Heidelberg, New York: 101-120.

Vink, A., 2004. Calcareous dinoflagellate cysts in South and equatorial Atlantic surface sediments: diversity, distribution, ecology and potential for palaeoenvironmental reconstruction. *Marine Micropaleontology*, **50**: 43-88.

Wade, B.S., Bown, P.R., 2005. Calcareous nannofossils in extreme environments: The Messinian Salinity Crisis, Polemi Basin, Cyprus. *Palaeogeography, Palaeoclimatology, Palaeoecology*, **233**(3-4): 271-286.

Wendler, I., Zonneveld, K. A. F., Willems, H., 2002a. Calcareous cyst-producing dinoflagellates: ecology and aspects of cyst preservation in a highly productive oceanic region. *In*: Clift, P.D., Kroon, D., Geadicke, C., Craig, J. (Eds.). The tectonic and climatic evolution of the Arabian Sea region. *Geological Society Special Publication*, **195**: 317-340.

Wendler, I., Zonneveld, K. A. F., Willems, H., 2002b. Oxygen availability effects on early diagenetic calcite dissolution in the Arabian Sea as inferred from calcareous dinoflagellate cysts. *Global and Planetary Change*, **34**: 219-239.

Wendler, I., Zonneveld, K. A. F., Willems, H., 2002c. Production of calcareous dinoflagellate cysts in response to monsoon forcing off Somalia: a sediment trap study. *Marine Micropaleontology*, **46**: 1-11.

Zonneveld, K.A. F., Brune, A.,Willems, H., 2000. Spatial distribution of calcareous dinoflagellates in surface sediments of the South Atlantic Ocean between 13°N and 36°S. *Review of Palaeobotany and Palynology*, **111**: 197-223.

Zonneveld, K.A.F., Versteegh, G.J.M., De Lange, G. J., 2001. Palaeoproductivity and post-depositional aerobic organic matter decay reflected by dinoflagellate cyst assemblages of the Eastern Mediterranean S1 sapropel. *Marine Geology*, **172**: 181-195.

Zonneveld, K.A.F., 2004. Potential use of stable oxygen isotope composition of *Thoracosphaera heimii* for upper water column (thermocline) temperature reconstruction. *Marine Micropaleontology*, **50**(3/4): 307-317.

Zonneveld, K.A.F., Meier, K.J.S., Esper, O., Siggelkow, D., Wendler, I., Willems, H., 2005.The (palaeo-) environmental significance of modern calcareous dinoflagellate cysts: a review. *Paläontologische Zeitschrift*, **79**(1): 61–77.

Chapter 5

Conclusions and implications

The results presented in this thesis are based on the first detailed calcareous dinoflagellate cyst records from the Neogene central and eastern Mediterranean realm related to the environmental changes caused by the MSC. For the first time the quantitative analysis of fossil calcareous dinoflagellates was performed by SEM analyses applying a new or modified method.

This study has shown that calcareous dinoflagellates reflect long- and short- term environmental changes in the Neogene Mediterranean. This shows their potential for the reconstruction of past marine environments. We demonstrated that individual species reflect changes in environmental factors of the surface water in particular temperature, nutrient concentrations, salinity and stratification. These in turn we attributed to variations in climatic factors such as arid-warm (*C. albatrosianum*) versus arid-cool (*T. heimii*) and humid-warm (e.g. *C. stella*) versus humid-cool (e.g. *L. granifera*) and in physical oceanographic factors such as dynamic of the circulation i.e. estuarine versus anti-estuarine which in turn is coupled with upwelling (*L. granifera*) versus downwelling and mixing (*P. tuberosa*) versus stratification (*C. albatrosianum*, *T. heimii*) which again controls the nutrient availability in the different water levels. High nutrient concentrations in the upper surface water usually associated with lower salinities are reflected in high percentages of *L. granifera* whereas high nutrient concentrations at depths around the DCM layer are reflected by *T. heimii*. In addition, the presence of neritic, coastal and open oceanic species enabled a nearshore – offshore differentiation and an estimation of the approximate depth of the setting or neritic influences.

The outcome of this study provides new insight both, in the spatial and temporal evolution of calcareous dinoflagellates during late Neogene time in the central and eastern Mediterranean area and to the palaeoenvironmental conditions preceding and following the MSC. Although numerous studies during the last decades involved in unravelling the history of the MSC several questions could not be answered in detail so far. For example, the evolution of the pre-evaporitic stage i.e. the transition from normal marine to highly restricted conditions is still under debate (e.g. Orszag-Sperber et al., 2009). Especially was it (e.g. salinity increase and shallowing) a progressive or a stepwise evolution and did it take place simultaneously in the different basins with similar consequences and which role played local differences (e.g. climate, continental runoff and tectonic).

Our studies revealed that significant palaeoenvironmental changes already occurred long before the onset of the evaporite formation. According to our results from both basins (Caltanissetta and Pissouri Basin) first expressions of restricted conditions are already apparent in the upper Tortonian (drop in total cyst numbers), although with a slight west-east time offset and a stronger magnitude in the central Caltanissetta basin. The following evolution up to the onset of the MSC in the Caltanissetta Basin is characterised by two phases shown by our data. In the first phase conditions were relative stable warm and oligotrophic (~7.24 – ~6.81 Ma) reflected by the dominance of *C. albatrosianum* but with fluctuations in total cyst abundances. A major drop of *C. albatrosianum* occurs at 7.17 Ma. The second phase is initiated by a drastic and abrupt change (at ~6.78 Ma) in the dinoflagellate cyst association. Total cyst abundances decreased on average, superimposed by drastic changes in species dominance, episodic exceptional peak occurrences (e.g. at ~6.52 Ma and ~6.47 Ma) and barren samples (i.e. ~6.67 Ma and ~6.63 Ma) up to the complete disappearance of calcareous dinoflagellates at ~6.0 Ma, about 40 kyrs before the onset of the MSC at 5.96 Ma. In general these changes reflect shifts in salinity, temperature and nutrient concentrations. Exceptional peak occurrences we interpreted as hints for episodic enhanced Atlantic water incursions.

In the eastern Mediterranean basin the evolution is comparable but less pronounced and with a more transitional character. According to our data the salinity is lower and nutrient concentrations are higher which was attributed to stronger continental runoff. In addition total cyst abundances were much lower in the central than in the eastern Mediterranean basin at least during the upper Tortonian, lower Messinian and early Pliocene. Probably the effect of the basin restriction was less pronounced in the eastern basin, so that scarce findings of calcareous dinoflagellates were made even up into the barre jaune unit shortly before the onset of the formation of the massive lower gypsum deposits of the MSC. Nevertheless, significant changes occurred at e.g. about 7.51, 7.1, 6.7 and 6.42 Ma, the latter one being the most drastic event. Thenceforth strong restricted and unstable conditions are reflected in the dinoflagellate cyst association.

Successive major palaeoenvironmental changes at some of these points have also been recognized by studies on benthic and planktonic foraminifera and on other microfossils in Mediterranean basins (e.g. Seidenkrantz et al., 2000; Blanc-Valleron et al., 2002; Kouwenhoven et al., 2006). Our results show that besides a general accordance in the major trends local differences existed in the hydrological and climatological conditions and in local tectonic activity. A stronger Atlantic influence (peak occurrences) is apparent in the central

Mediterranean basin which resulted in stronger and more frequent shifts of the surface water temperature. A trend of increasing nutrient concentrations and decreasing temperatures at least temporally toward the MSC is visible in both basins as indicated primarily by *L. granifera* and *C. albatrosianum* can be interpreted as a change towards a more humid and cooler climate associated with enhanced continental runoff.

At the beginning of the Pliocene (re-establishment of fully marine conditions) strong eutrophication is reflected in both basins by a significant abundance peak of *L. granifera*. This we interpreted as caused by a combination of increased continental runoff and local upwelling. Hence, we suggest that during the early Pliocene upwelling and fluvial input caused the high productivity of *L. granifera*.

After a transitional period of about 100 kyrs, dominated by *L. granifera*, *C. albatrosianum* took over the supremacy again indicating a renewed change in the hydrological and climatic conditions, from eutrophic and somewhat cooler surface waters (humid and cool climate) towards warmer and oligotrophic conditions (warm and arid climate). This final change probably was associated with a change from an estuarine to anti-estuarine system (nutrient loss) as it prevails today in the Mediterranean Sea. Nevertheless, the Pliocene calcareous dinoflagellate association still significantly differs from that of today particularly with respect to species dominances. Consequently, it can be assumed that the Pliocene Mediterranean system also was different from the present one.

We have shown that significant palaeoenvironmental changes occurred in successive steps since the late Tortonian and that increasing restriction is well reflected in the calcareous dinoflagellate association but no indication for a continuous increase in surface water salinity and significant decrease in water depth was observed. However, the decrease in neritic species toward the MSC we related to hostile bottom water conditions possibly as a result of early brine formation.

Future perspectives

This study has shown that Neogene calcareous dinoflagellates respond reliable and sensitive to known and supposed changes in the surface water conditions providing important specific information about the environmental and climatic changes in the time preceding and following the MSC in the central and eastern Mediterranean basins. Therefore with our results we proofed the usefulness of fossil calcareous dinoflagellates for palaeoenvironmental reconstructions. Nevertheless, future (palaeo)environmental studies on calcareous dinoflagellates will surely result in a refinement of the existing approaches and methods and improve the data quality and thus the palaeoenvironmental interpretation. For future research on fossil calcareous dinoflagellates we make the following recommendations.

Investigation of other localities

To increase our understanding of the spatial and temporal evolution of the calcareous dinoflagellates and to contribute to a better understanding of the MSC it is necessary to extend our studies to additional sections across the Mediterranean to complete the West–East transect as it was planed in our initial project approach. For several reasons we had to deviate from the original project so that it was not possible to include the western localities i.e. southern Spain and north-west Morocco and northern Italy.

The sections of Morocco and Spain are located in key areas of the two main late Miocene Atlantic-Mediterranean gateways (the Rifian corridor to the south and the Betic corridor to the north) and thus can provide valuable data about the hydrological conditions in the most western part of the Mediterranean realm and of the Atlantic side gained from calcareous dinoflagellate records. The inclusion of localities from the Apennine foreland basins which represent the northward extension of the central Mediterranean Sicily basins in addition would enable the comparison and differentiation of local and Mediterranean-wide effects. Of all localities we already carried out first pilot studies. First results will be briefly presented in the following.

Morocco (Bou Regreg area)

The Neogene deposits of the Bou Regreg area comprise a complete open marine (500 - 700 m depth) Atlantic parallel section of the MSC that covers the upper Tortonian to late Pliocene with the classic blue marls (Krijgsmann et al, 1999; Van der Laan et. al, 2005). The blue marls of the Bou Regreg area are well investigated and are therefore very well suited as the Atlantic reference profile for the Mediterranean deposits.

In total we took 70 samples from the Miocene/Pliocene transition. The samples were taken by a common field work with our colleagues from the Netherlands. In general 8 to 10 samples were taken per cycle (precession cycle) which corresponds to a 2 - 3 ka resolution. Off the 27 samples from the upper Tortonian/upper Messinian interval 562 cysts have been counted including about 24 species. *C. albatrosianum* significantly dominates the Tortonian association with up to 100%. The upper Messinian is characterised by a significant decrease in *C. albatrosianum* and strong fluctuations in relative abundances. The percentage of *L. granifera* and *P. edgarii* notably increased indicating a raise in the nutrient level of the surface water associated with a shallowing of the setting.

Spain (Sorbas Basin)

One of the most complete Messinian sedimentary sequences of the westernmost Mediterranean is located in the Sorbas Basin (SE Spain) forming the northern connection (Betic corridor) between the Atlantic and Mediterranean during the Late Miocene (Krijgsman et al., 2001). The Messinian deposits of the Sorbas basin are excellently exposed and astronomically calibrated (Krijgsman et al., 2001) and thus provide an excellent basis for studies on calcareous dinoflagellates. The studies here were planed to focus on the alternation between sapropels and marls.

From the four analyzed samples of the upper Messinian a total of 957 dinoflagellate cysts were counted and eight species distinguished. The by far dominating species are *C. albatrosianum* (55%) and *L. granifera* (35%). No consistent relationship between the lithological units and the two species was evident. Interestingly the oligotrophic warm-water species *C. albatrosianum* distinctly dominates two of the sapropel layers with up to 83% whereas *L. granifera* an indicator for high nutrient concentrations reached its highest abundance (42%) within the marl layer. Future studies could unravel this phenomenon and provide important information of the surface water conditions during sapropel and marl formation. These findings confirm our observations from the Caltanissetta Basin (see Chapter 4).

North Italy (Monte del Casino Section)

The Monte del Casino section is located in northern Italy. The sedimentary sequence represents a deep foreland basin setting of the northern Apennines with estimated water depth of more than 1000 m (Kouwenhoven, 2000). Sixteen samples have been analyzed so far from the upper Tortonian/lower Messinian interval.

Off the 954 cysts counted, about 17 species were identified. *C. albatrosianum* clearly dominates the Upper Tortonian (88%) and lower Messinian assemblages (76%) on average, although with a notable decrease during the Messinian. A significant drop in the upper Tortonian temporarily interrupted its dominant position. A concurrent increase of *L. granifera* and *P. edgarii* indicate increased nutrient conditions and a short-time shallowing of the basin.

The first results of the three Neogene land section confirm our results and major trends in the dinoflagellate cyst association recorded from the Sicily and Cyprus sections. *C. albatrosianum* dominates the Tortonian/Messinian assemblages but shows an upward trend to decrease. However, slight differences in the associations are also visible pointing to local differences. A continuation of these studies seems to be very promising.

Method

Advances in SEM-technology and software in future works will contribute to a significant improvement in the use of SEM microscopy for micropalaeontological studies. The combination of Electron backscatter diffraction (EBSD) with the SEM enables the determination of the orientation of the crystallographic c-axis under the SEM (Kameda et al., 2005). Software improvement will allow the offline processing (counting and identification) and makes the data analysis more independent from SEM sessions (schedules) and thus will increase the final sample throughput. Of course, there is certainly still room for improvement in sample preparation and equipment particularly in respect to a higher exploitation of specimen of rather unproductive periods such as during the upper Messinian. This would probably not change the general trend but at least confirm it.

Stratigraphic extension

Extension of the dinoflagellate cyst dataset through the full succession of the Pliocene Trubi marls from Cyprus, Sicily and from Atlantic Morocco would complete our knowledge about the calcareous dinoflagellate evolution in the Mediterranean realm and surely may provide additional important information about the development of the Mediterranean hydrology. Our Pliocene dinoflagellate associations still significantly differ from the present Mediterranean ones in composition as well as in the clear present east-west division. So far it could not be detected when these associations established. Further investigations could fill this gab and in addition would provide important information about the Mediterranean system (e.g. strong seasonality) as the distribution of calcareous dinoflagellates is closely related to the physical parameters in the surface waters.

Studies on core material

Core material from the Mediterranean deep basins would enable high resolution sampling on undisturbed sedimentary records related to the MSC and to subsequent strata and thus offering ideal conditions to further strengthen the proxy value of fossil calcareous dinoflagellates.

Studies on recent material (sediment traps/water samples)

Sediment trap and water sample studies throughout the Mediterranean Sea would provide real-time information about the present distribution of calcareous dinoflagellates and their environmental relations especially to species specific depth habitats and to possible seasonality in cyst production. Comparison with the underlying surface sample may reveal valuable information about lateral transport and possible (selective) degradation processes (e.g. Dale, 1992; Wendler et al., 2002b).

Comparison with other proxies

Combined studies with terrestrial proxies such as pollen would provide information on (local) continental climate for comparison with surface water conditions. This would be especially interesting in respect to times and origin of sapropel formation which are still contentious and unresolved (Krijgsman et al., 2002b). Our results have shown that the climatic implication and suggestions for times of sapropel formation (humid/warm) do not necessarily coincide with popular conceptions especially in respect to the prevailing surface water conditions (nutrient enriched, warm, humid climate) (e.g. Rossignol-Strick, 1983). It could not be verified that sapropels tend to exhibit higher cyst concentrations (productivity) than marls and that surface water conditions are generally nutrient enriched. We have shown that sapropels formed under both, oligotrophic/warm and oligotrophic/cool stratified surface water conditions. Oligotrophic and stratified surface water conditions during sapropel formation have been also suggested by Kemp et al., (1999) and Emeis et al. (2000).

An addition it would be interesting to combine the environmental signals gained from the distribution pattern with isotope data derived from calcareous dinoflagellates. Currently at the University of Bremen the proxy potential for SST was tested on *T. heimii* (Kohn, 2009). Future research has to proof if this approach is also practicable for fossil material (e.g.

Tertiary) and how far back in time it is applicable. As we have shown in this work the scarcity of *T. heimii* in past environments, at least up to the Early Pliocene could limit its applicability to Quaternary studies since of the presently required carbonate volume for isotope measurements (Karwath, 1999).

Intraspecific morphological variations

In future studies more attention should also be drawn on intraspecific morphological cyst variations and possible relations to specific environmental parameters (e.g. salinity). This approach has unfortunately missed out in this study a little bit although morphological variations were realized but no significant trends. This would imply a more detailed consideration of various factors.

References

Blanc-Valleron, M.-M., Pierre, C., Caulet, J.P., Caruso, A., Rouchy, J.-M., Cespuglio, G., Sprovieri, R., Pestrea, S., Di Stefano, E., 2002. Sedimentary, stable isotope and micropaleontological records of paleoceanographic change in the Messinian Tripoli Formation (Sicily, Italy). *Palaeogeography, Palaeoclimatology, Palaeoecology,* **185**: 255- 286.

Dale, B., 1992b. Thoracosphaerids: Pelagic Fluxes. In: S. Honjo (Ed.). Dinoflagellate Contributions to the Deep Sea. *Ocean Biocoenosis Ser*ies, Oceanographic Instituion, Woods Hole, Massachusetts, **5**: 33-44.

Emeis, K.-C., Sakamoto, T., Wehausen, R., Brumsack, H.-J., 2000. The sapropel record of the Eastern Mediterranean Sea – results of Ocean Drilling Program Leg 160. *Palaeogeography, Palaeoclimatology, Palaeoecology*, **158**: 371–395.

Kameda, J., Yamagishi, A., and Kogure, T., 2005. Morphological characteristics of ordered kaolinite: investigation using electron back-scattered diffraction. *American Mineralogist*, **90**: 1462–1465.

Kemp, A.E.S. Pearce, R.B. Koizumi, I. Pike J., Rence J. 1999. Role of mat-forming diatoms in the formation of Mediterranean sapropels. *Nature*, **398**: 57-61.

Karwath, B., 1999. Ecological studies on living and fossil calcareous dinoflagellates of the equatorial and tropical Atlantic Ocean. Berichte Fachbereich Geowissenschaften, Universität Bremen, **122**: 175pp.

Kohn, 2009. The stable oxygen isotope signal of the calcareous-walled dinoflagellate *Thoracosphaera heimii* as a new proxy for sea-surface temperature. *Dissertation zur Erlangung des Doktorgrades der Naturwissenschaften*, Universität Bremen: 117pp.

Kouwenhoven, T. J., 2000. Survival under stress: benthic foraminiferal patterns and Cenozoic biotic crises. *Geologica Utraiectina*, **186**: 1-206.

Kouwenhoven, T. J., Morigi, C., Negri, A., Giunta, S., Krijgsman, W., Rouchy, J.-M., 2006. Paleoenvironmental evolution of the eastern Mediterranean during the Messinian: Constraints from integrated microfossil data of the Pissouri Basin (Cyprus). *Marine Micropaleontology*, **60**(1): 17-44.

Krijgsman, W., Langereis, C.G., Zachariasse, W.J., Boccaletti, M., Moratti, G., Gelati, R., Iaccarinof, S., Papani, G., Villa, G., 1999. Late Neogene evolution of the Taza–Guercif Basin (Rifian Corridor,Morocco) and implications for the Messinian salinity crisis. *Marine Geology*, **153**:147–160.

Krijgsman, W., 2002. The Mediterranean: Mare Nostrum of Earth sciences. *Earth and Planetary Science Letters*, **205**: 1-12.

Rossignol-Strick , M., 1983. African monsoons, an immediate climate response to orbital insolation. *Nature*, **304**: 46–49.

Seidenkrantz, M.-S., Kouwenhoven, T.J., Jorissen, F.J. Shackleton, N.J., van der Zwaan, G.J., 2000. Benthic foraminifera as indicators of changing Mediterranean–Atlantic water exchange in the late Miocene. Marine Geology, **163**: 387–407.

Van der Laan, E., 2005. Regional climate and glacial control on high-resolution oxygen isotope records from Ain el Beida (latest Miocene, northwest Morocco): A cyclostratigraphic analysis in the depth and time domain. *Paleoceanography*, **20**: 1-22.

Wendler, I., Zonneveld, K. A. F., Willems, H., 2002b. Oxygen availability effects on early diagenetic calcite dissolution in the Arabian Sea as inferred from calcareous dinoflagellate cysts. *Global and Planetary Change*, **34**: 219-239.

Acknowledgements

I am indebted to many people for their long-lasting support and encouragement which was invaluable for the successful completion of this work. In the following lines some of them are gratefully acknowledged. However, I am aware of the fact that there are many more and these words cannot express the gratitude for all of those.

First of all, my most sincere thanks go to Prof. Dr. Helmut Willems who gave me the opportunity to undertake this PhD project. All of this work would not have been possible without his general support and encouragement throughout the entire duration of this work. I really enjoyed our fieldtrip to Morocco and Spain, where I had the opportunity to be part of an unlimited enthusiasm and working spirit of *a geologist in the field.* We really had a great time, together with our colleagues from the Netherlands, Dr. Erik Snel and Dr. Erwin van der Laan. Many thanks also go to Prof. Dr. Rüdiger Henrich – another enthusiastic *geologist in the field*, for his kind agreement to provide the second review of this thesis. I am also grateful to Dr. Gerard Versteegh for his guidance, support, helpful suggestions and good advice, throughout the entire project.

I sincerely thank our cooperation partners from the Netherlands and from France, for providing the sample material, discussions and the reviews of the manuscripts, especially to Dr. Frits Hilgen, Prof. Dr. Jean-Marie Rouchy and Prof. Dr. Fabienne Orszag-Sperber. My special thanks go to Dr. Tanja Kouwenhoven, Dr. Erik Snel and Dr. Erwin van der Laan, for their constant assistance with the sample procurement. Erik and Erwin, it was a great pleasure to have your company during the field works in Morocco. I really enjoyed sample drilling, extensive discussions in the field and elsewhere. Prof. Dr. Nadja Barhoun is thanked for her help in obtaining the working permits and general assistance during our field trip in Morocco.

Also, I thank all the gurus of the Messinian Salinity Crisis which I had the great pleasure to meet on some conferences and joined field trips in Italy (Parma and Urbino), France and elsewhere, namely e.g. M.-B. Cita, S.-M. Iaccarino, F. Orszag-Sperber, J.-M. Rouchy, F. Hilgen, W. Krijgsman, M. Roveri, V. Manzi and the gurus of the calcareous dinoflagellates e.g. H. Keupp and R. Kohring, besides many others.

Many thanks to all members and former members of the Division of Historical Geology and Palaeontology who contributed to a positive atmosphere, provided general support and shared many interesting discussions, scientific once and beyond: Angelika, Anne, Annemiek, Carmen, Cheng Liang, Christiane, Dorit, Dorothee, Erna, Ewa, Gabriele, Gerard, Hartmut, Helmut, Ines, Ilham, Jacob, Jan-Peter, Kai-Uwe, Kara, Karin, Konstantin, Chen Liang, Marion K., Maria M., Matthias, Mirja, Monika Ki., Monika Ko., Nicole, Oliver, Petra, Qinghai, Rehab, Sonja, Stefanie, Stijn, Sven, Thomas, Ulrike and Vanessa. Many thanks, to Jacob Nsiah for preparing the samples and to Hartmut Mai and

Petra Witte for assistance and advice for the SEM analysis. Karin Zonneveld I thank for general and expert advice.

Many thanks also to Erna, Maria and Marion M. for their permanent kindness and helpfulness, not only with administrative matters. Erna, I really enjoyed our Tai-Chi sessions in the Park together with my husband.

Special thanks go to my office and floor mates Cheng Liang, Dorit, Ewa, Ilham, Kara, Marion, Mirja, Sonja, Stefanie, Ulrike and Vanessa for the cheerful company and fruitful discussions during our famous tea-breaks. In addition I am grateful to Kara for careful reading and English corrections. The Sushi-group, especially Marion, Oliver, Sonja, Uli, Vanessa and partners, I thank for the pleasant hours outside the University.

My special thanks go to Ewa Susek, who was my roommate and good friend for more than three years and to Vanessa Lüer for her encouragement and friendship throughout the years. My sincere thanks go to many friends outside the University for their general interest in my work and for many welcome distractions, for example a relaxed dive in our lakes in and around Bremen.

This work was funded by the University of Bremen and the German Research Foundation (DFG) which enabled the realization of this project, the fieldtrip to Morocco and Spain participation in conferences and workshops.

Last but by no means least, I like to thank my family, especially my beloved parents for their endless love, patience, and continuous emotional support, they have always been there for me throughout my life. *Najdroższa mama i papa, dziękuję!*

Very special thanks also go to my siblings Willi, Brigitte, Ramona and relatives Danny, Patrick, Jessica, Justin, Thomas, Peter.

My beloved sun Marco and my beloved husband Jörg I especially thank for their neverending love, patience, thoughtfulness, understanding and support throughout all the years.

Für meinen geliebten Vater

Walter Bison
*13. 07. 1936 - 14. 07. 2008†

"Deine starken Hände, ... Unverdrossener als die meisten Haben sie geschafft, ... Und er lacht Leise, dass er sie nicht wecke." (nach H. Hesse)

MIX
Papier aus verantwortungsvollen Quellen
Paper from responsible sources
FSC® C105338

Printed by Books on Demand GmbH, Norderstedt / Germany